CW00724214

ATOMISTIC INTUITIONS

SUNY series in Contemporary French Thought
—————————————————————————————
David Pettigrew and François Raffoul, editors

ATOMISTIC INTUITIONS
An Essay on Classification

GASTON BACHELARD

Translated and with an introduction by ROCH C. SMITH

Preface to the French edition by DANIEL PARROCHIA

Originally published, in the French, as *Les intuitions atomistiques (essai de classification)* Deuxième édition revue et corrigée. Préface de Daniel Parrochia. © Librairie Philosophique J. Vrin, Paris, 1975; 2015. http://www.vrin.fr

Published by State University of New York Press, Albany

For information, contact State University of New York Press, Albany, NY
www.sunypress.edu

Library of Congress Cataloging-in-Publication Data

Names: Bachelard, Gaston, 1884-1962, author.
Title: Atomistic intuitions : an essay on classification / by Gaston Bachelard ; translated by Roch C. Smith ; preface by Daniel Parrochia.
Other titles: Intuitions atomistiques. English
Description: Albany : State University of New York, 2018. | Series: SUNY series in contemporary French thought | Includes bibliographical references and index.
Identifiers: LCCN 2017053072| ISBN 9781438471273 (hardcover : alk. paper) | ISBN 9781438471297 (e-book)
Subjects: LCSH: Atomism.
Classification: LCC BD646 .B313 2018 | DDC 146/.5—dc23 LC record available at https://lccn.loc.gov/2017053072

10 9 8 7 6 5 4 3 2 1

CONTENTS

TRANSLATOR'S INTRODUCTION

ATOMISTIC INTUITIONS IS THE SIXTH OF TWENTY-THREE BOOKS PUBLISHED by Gaston Bachelard during his lifetime. Of these, fully one-half deal with the epistemology of science, while the remainder engage with broader philosophical issues from a variety of perspectives, with special emphasis on the literary imagination.[1] With the publication of this translation, sixteen of Bachelard's books, including an unfinished posthumous work, are now available in English. Yet only a quarter of these are translations of works on science. Clearly, more needs to be done, and it is my hope that the present volume will encourage more translations and will further scholarly efforts to make Bachelard's significant explorations of modern scientific thought more widely recognized in the English-speaking world.

Bachelard, who died in 1962—the same year Thomas Kuhn published his *Structure of Scientific Revolutions*—had been investigating and explaining what he would call *The New Scientific Spirit* for some thirty years.[2] It is not my purpose here to examine in detail the remarkable parallels between Kuhn's notion of the "paradigm shift" particularly evident in modern science and Bachelard's own observations on the revolution in contemporary scientific thought. Suffice it to point out that there are significant similarities in outlook and conclusions, despite differences in presentation and level of detail between the two. Briefly put, among the differences, Kuhn focuses especially on the scientific process as it is revealed in the applied practice of science, both over time and in the present. While not neglecting such practice, Bachelard is more likely to call attention to the history of science for the lessons it teaches on past errors, and to the philosophical implications of contemporary scientific practice. Examples of congruity between these kindred spirits include what Kuhn calls "anomaly" and Bachelard identifies as "epistemological obstacles" at work in the revolutionary process of scientific discovery.[3] Similarly, both are led to consider the pedagogical implications of such a method for students who are too often introduced to science as accumulated knowledge rather than as a process of perpetual discovery. Kuhn points out that "the textbook tendency to make the development of science linear hides a process that lies at the heart of the most significant episodes of scientific development."[4] Bachelard decries the secondary school practices in the France of his day, where, in abandoning problem-solving in favor of teaching summary knowledge, physics and chemistry serve to "misunderstand the real meaning of the scientific mind."[5] Such pedagogical considerations are perceptively explored by Cristina Chimisso, who sees pedagogy as an underlying and distinctive concern in Bachelard, going well beyond instances of instructional shortcomings. In a recent book Chimisso's well-researched historical context supports the argument that Bachelard's "epistemology and his pedagogy were not separate: rather, his epistemology shaped his pedagogy, and his pedagogy inspired his conception of science."[6] Space does not permit a fuller exploration of correspondences between Kuhn and Bachelard on this and other questions, but it certainly would not be the first time that two thinkers, writing on issues of their time, came separately to similar conclusions. Yet the fact that Bachelard's work precedes that of Kuhn by three decades would suggest, at the very least, the need for more attention to continental philosophy by English-speaking readers. Kuhn himself, of course, in his transition from theoretical physics to the history and philosophy of science, spanned that partition and came to know the world of European philosophy well, including figures such as Alexandre Koyré, Émile Meyerson, and Hélène Metzger, who serve as points of reference for Bachelard's own discussions.[7]

Recent years have seen efforts to bridge this philosophical divide by offering several analyses of Bachelard to readers of English.[8] Among these recent considerations of Bachelard's work, one, in particular, calls attention to the role of atomism in Bachelard. In a chapter-length appendix to a book that examines Bachelard's view of mathematical rationalism as a creator of realities in science, Zbigniew Kotowicz suggests that "Bachelard's conception of time is atomist, and he thinks like an atomist."[9] Especially intriguing is the second part of this proposition, since it extends Bachelard's recognizable temporal atomism in *Intuition of the Instant* (1932) to the rest of his thought. *Atomistic Intuitions*, published the year after Bachelard's book on the instant emerges as an obvious instance of his atomistic thinking. But Kotowicz reaches well beyond this early essay to include works from what I have called Bachelard's "epistemological trilogy."[10] Arguing that, for Bachelard, "the ontologising power of mathematics comes into full view and is put into effect in an atomist universe, [and that] ... science is discontinuous as a consequence of the atomist nature of our rationality,"[11] Kotowicz points the way to a fuller consideration of atomism in Bachelard. Although Kotowicz considers *Atomistic Intuitions* to be largely dismissive of traditional atomism—what he calls "the concept of the atom as a bit of material substance"—he does suggest that "it was atomism that gave [Bachelard] an ontology of discontinuity, of perfect mobility, in which he could articulate his project."[12] But the question remains as to whether it was atomistic thinking—"the atomist nature of our rationality"—that led Bachelard to a perception and ultimate understanding of the new science, as Kotowicz proposes, or whether Bachelard's open-minded investigation and understanding of that very science led, in turn, to his notions and appreciation of atomism, both in science and elsewhere. This question, and undoubtedly others surrounding the notion of atomism in Bachelard, warrant further inquiry. *Atomistic Intuitions*, a work that, under the broad umbrella of atomism, explores the encounter between realism and idealism, between the empirical and the axiomatic, and, in so doing, uncovers the "eclecticism" (98) of philosophical atomism, provides an indispensable starting point for just such a survey.

Even prior to the most recent twenty-first-century studies, a few English-speaking voices had already been addressing Bachelard's epistemology over the last two decades of the twentieth century. In addition to my own overview of Bachelard's writings on science, as part of an introduction in 1982 to the full spectrum of his thought,[13] there were notable contributions by Mary Tiles, whose thorough analysis of Bachelard's philosophy of science appeared in 1984,[14] and by Mary McAllester, whose 1989 edited collection of essays and 1991 book of translations and commentary include several considerations of Bachelard's epistemology.[15] Indeed, McAllester, who, along with Eileen Rizo-Patron, has translated Bachelard's philosophical musings on time,[16] has also provided an English translation for a key work of Bachelard's philosophy of science, namely, his 1938 *La formation de l'esprit scientifique*.[17]

Although *Atomistic Intuitions* comes across as less technical and more accessible than his previous epistemological writings, Bachelard continues to have the philosopher, and particularly the philosopher of science, in mind as his reader. But given Bachelard's later work on the imagination, now widely available in English, this volume will also be of interest to scholars in the humanities, especially those focusing on literature. This, after all, is the first work in which Bachelard, in a chapter on the metaphysics of dust, explores (albeit skeptically) the fascinating realism of matter. Bachelard will discuss the hazards of these realist attractions, these "epistemological obstacles," in such works as the 1938 *Formation of the Scientific Mind*, and *The Psychoanalysis of Fire*, as well as in *The Philosophy of No*, published two years later. Indeed, his 1938 book on the allure of fire, where Bachelard allows himself to be captivated by the enchanting images of that "element," will open the gates to an entire world of poetic imagery in his subsequent works on the material imagination and its phenomenology. But, more fundamentally, and perhaps more significantly, it is by furthering our appreciation of the philosophical implications Bachelard draws from contemporary science that *Atomistic Intuitions* contributes to a fuller comprehension of Bachelard's overall writings. As I have argued elsewhere, the perspectives and the intuitive *esprit de finesse* that Bachelard garnered from science, and that are evident in his epistemological work, provide a unique, indeed an essential key to understanding his extensive writings on the imagination.[18] Given such reasons for a broadened interest in this book, I have included explanatory notes that may not have been necessary for Bachelard's originally intended reader or present-day philosophers of science, but that may prove helpful to a wider readership. All such notes, as well as the clarifying material I have added to Bachelard's own notes, are kept in brackets.

Bachelard sets out in this book to explore philosophical doctrines of the atom over time, from Democritus to early twentieth-century physics, including, in particular, the atomistic intuitions such doctrines represent.[19] As his subtitle indicates, he proposes a classification of such intuitions as they are transformed over the course of history. But more than a mere taxonomy, this is a voyage of philosophical exploration, with extended stays along the way to an ultimate destination: what Bachelard calls the "new world" of microphysics and the "metamicrophysics" it calls forth (9). Each of Bachelard's several intervals along his journey to axiomatic atomism—including the atomisms of antiquity, realism and related issues of atomic composition, positivism, and Kantian criticism—offers its own intrinsic interest and is worth a protracted layover. But all take on their fullest significance as steps that must be traversed, and in certain ways rejected, en route to modern atomistics, the axiomatic science of atomic physics. Indeed, for readers of English, words like "atomistics" or "atomic science," when applied to the scientific revolution of the twentieth century, are often more clearly rendered, as I have occasionally done, simply as "atomic physics."

Bachelard's goal in this work is to develop a metaphysical context for modern atomistic science not only by reviewing its philosophical history but by drawing conclusions from how that science operates. His exploration of atomistic intuitions over time leads to the observation that "modern atomic physics gives us a brilliant example of axiomatic thought. It teaches us to think about the details of atomic being as analytically independent and subsequently to show their dependence on synthesis" (96). Analysis belongs to the realism and positivism of prior scientific thinking. Such outlooks have not been abandoned altogether by modern microphysics, but they have been transcended and incorporated into a science of postulates and axioms that relies on rational synthesis for its discoveries. As Bachelard points out, "it would thus be an error henceforth to consider atomistics as the analytical study of a fundamental element found at the origin of an intuition. Atomistics is, on the contrary, a wholly synthetic construction that must rely on a *body* of assumptions. That is why contemporary atomism is truly productive atomism. This atomism owes its productivity to the *compound character of the simple atom*" (92). For modern physics, the atom is a construct whose building blocks are mathematical, hence synthetic rather than analytical. A related narrative that helps explain the transformation of atomistic intuitions over time has to do with the gradual abandonment of the notion of the atom as stable and occupying space, however small, to one that sees the atom as dynamic. With its reliance on mathematical relations, the atom finds its objectivity in the dynamic synthesis of rational concurrence, rather than in any physical object.

In the case of chemistry, Bachelard reminds us, intuitions, in going beyond positivist analysis, could be productive of scientific understanding. His exploration of positivism in chapter 4 traces the struggle between positivism's emphasis on the phenomena—with its particular attention to atomic weights (though not atoms)—and intuition-inspired theory. Where positivism sought to move "from phenomena to principles," John Dalton and others went "from principles to phenomena, from intuition to experimentation" (57). As Bachelard's review of the contributions of major nineteenth-century chemists makes clear, atomistic intuitions prevailed as, "very rapidly, the chemical phenomenon was partitioned arithmetically and conformed readily to the atomic hypothesis" (62) where "despite every prohibition, the life of first intuitions subsisted" (65).

Bachelard's own experience as a secondary school teacher of physics and chemistry a decade prior to the publication of *Atomistic Intuitions* gives particular poignancy to his description of this encounter. He had employed the textbooks and he had endeavored to teach an official curriculum so firmly grounded in positivism that it largely avoided, through circumlocution and deferral, any reference to atomism. As Bachelard explains: "It took skill not to utter the word *atom*. One always thought about it, but one never talked about it" (56). It is just this sort of experience that will encourage Bachelard to consider not only philosophical issues, but also pedagogical ones in his later work.[20] The implications of the scientific revolution in both chemistry and physics for the teaching of science will remain a concern for Bachelard as he demonstrates the "applied rationalism" or "rational materialism" of contemporary science.[21]

Bachelard's search for a fuller understanding of the revolution in twentieth-century science combines, under the umbrella of atomism and its intuitions, previous analyses from two separately published studies on the transformations occurring in physics and chemistry. Four years prior to the publication of *Atomistic Intuitions*, Bachelard had analyzed the conceptual shift brought about by the theory of relativity in *La valeur inductive de la relativité*.[22] This work was followed three years later by an exploration of the development of chemical theory during the nineteenth century in *Le pluralisme cohérent de la chimie moderne*.[23] Clearly, the atom and how it is conceived over time, both intuitively and conceptually, emerges as a key to our comprehension of the modern scientific revolution. *La valeur inductive de la relativité* examines the contemporary terminus of that voyage of understanding. In this early work underscoring the differences between Newtonian and Einsteinian mechanics, Bachelard, anticipating Kuhn, advances the idea of a "rupture between the two methods ... that follow ... two entirely heterogeneous orders of thought."[24] More importantly, he carefully sets out to demonstrate how, in the Einsteinian system, that difference stems from the application of mathematics, in particular tensor calculus, to construct a potential reality, rather than to discover it directly, as had traditionally been the case. In the world of relativity, Bachelard points out in one of several such aphorisms, "the real is demonstrated, not revealed."[25] Rather than the deductive analysis of traditional mechanics, such a mathematically constructive conquest proceeds through a process of induction and synthesis. As Bachelard concludes, relativity is "touched by the light of axiomatics"[26] in a science that "has increased the freedom of postulates while simultaneously calling for an entirely new availability of intuition."[27] Bachelard seems to have determined that, in order for this transformation of intuitions in axiomatic science to be most fully appreciated, the changes in their function are best explained historically. Here, once again, we find an important purpose of the classificatory excursion that is *Atomistic Intuitions*. By means of classification of the historical, Bachelard aims at explanation of the contemporary.

However, the role of atomistic intuitions as well as the relevant rational abstractions develop differently in chemistry than in physics. Not only does Bachelard consider these differences in chapter 3 on the composition of phenomena and chapter 4 on positivism, he had also addressed them in *Le pluralisme cohérent de la chimie moderne*. A look at the earlier work, with its focus on chemistry, reveals some of the roots of Bachelard's profound interest in the relation of reason and matter. The source of the coherent pluralism of modern chemistry is, of course, the periodic table and the remarkable capacity of this ordinal scheme, especially as it develops, not just to classify the diversity of observable phenomena but to introduce previously unknown elements. Although the problems and approaches are initially quite different from those of physics, here too it is not immediate reality but mathematical abstraction that introduces new possibilities and thus leads back, albeit indirectly, to the real. As Bachelard points out, "The problem of classification governs the problem of knowing a particular substance.... New substances do not correspond to entities found through observation.... They are *concepts made real*."[28] This "rational materialism" becomes even more pronounced as chemistry penetrates the atom to focus on the electron. As Bachelard indicates, "with the electron, the scientific explanation has, in effect, *gone beyond realism*."[29] The atomic number itself, in its correspondence with the positive charge of the nucleus, completes "the clear arithmetization of the periodic classification."[30] Chemistry now approaches the synthesis found in physics as "the rationalism of the possible has preceded and prepared the rationalization of the real."[31] The power of classification in chemistry is unique, of course, yet it shares with physics the constructive value of mathematical reason.

Many of the ideas on atomism explored so meticulously in this book thus have their origins in Bachelard's considerations of contemporary scientific revolutions in the works on chemistry and relativity that appeared in the few years prior to *Atomistic Intuitions*. In addition, an early examination of the metaphysical implications of contemporary physics, so central to Bachelard's discussion of atomism, can be found in a short essay originally published in the 1931–1932 issue of *Recherches philosophiques* and helpfully reissued as a chapter of *Études*, a posthumous collection of Bachelard essays presented by his well-known student and highly respected philosopher in his own right Georges Canguilhem.[32] These short pages, reviewing what Bachelard calls here the "noumenal" as opposed to the phenomenal character of mathematical constructions in microphysics, distill Bachelard's fundamental proposition that the contemporary physics of the atom introduces the possibility of a metaphysics for our time.

Bachelard's classificatory journey thus does more than merely examine former intuitions of the atom only to reject or avoid them. For his voyage of discovery reveals a conciliation of sorts among the varied philosophical attitudes concerning the atom. As he puts it: "We would therefore be right not to neglect any of the philosophical paths that I have endeavored to retrace in the course of this work. We should even find a way to establish correspondences among the various philosophies in order to succeed in truly submitting the atom to *thought*" (98). Atomism offers "a fusion of the principles of [Kantian] critical atomism and the teachings of realist atomism [that] is metaphysically impure," explains Bachelard (99) and is thus frequently ignored by metaphysical doctrines. But from the point of view of someone, like Bachelard, who is seeking a fuller understanding of the relation of contemporary science and philosophical atomism, "an initial fusion is necessary between idealist and realist theses," for, as he concludes, if such a fusion could be achieved, "we would be better prepared than we might think to follow the development of modern scientific atomism" (100). Understanding modern scientific atomism through the informed and thoughtful classification of intuitions of the atom thus remains an ongoing objective of this book, rather than its settled accomplishment. In keeping with Bachelard's pedagogical perspective, *Atomistic Intuitions* functions as a book of aspirations. Readers are invited to apply its lessons to their own attempts to come to terms with the still developing revolution in contemporary science.

ANY TRANSLATOR OF BACHELARD WILL FACE A FEW INTERESTING QUESTIONS stemming from the topic being investigated as well as from Bachelard's own linguistic propensities. Although such issues are less widespread in *Atomistic Intuitions* than in many of his other works, it may be helpful to highlight just a few. We have already seen, with the term "metamicrophysics," Bachelard's fondness for neologisms, especially when such expressions encapsulate a protracted alternative. In this case, the coinage summarizes the argument that contemporary atomistic intuitions offer a philosophical consideration of microphysics comparable to the metaphysics that, since Aristotle, has followed or gone beyond physics. Similarly, Bachelard's French term *chosiste*, which I have rendered here as "thing-oriented" (31), not only captures the emphasis on the physicality of one form of realist atomism but, at the same time, distinguishes it from the scientific "object" that results from synthesis. Related to his use of some neologisms in this work, Bachelard will occasionally engage in wordplay, as when he proposes that axiomatic atomism "*multiplied* the evidence" by "*adding* together" objective clarity and subjective certainty (89–90, my emphasis), or concludes that various schools of atomistic thought "mix a little a priori with a lot of a posteriori" (99).

Beyond neologisms and occasional wordplay, it may also be useful to call attention to the special use of certain terms. As I have pointed out in an endnote to chapter 3 (note 15), Bachelard uses the term "phenomenology" in this book to refer to the phenomena of science and not to the Husserlian philosophy manifested in some of his later works on the imagination. Words like *esprit* or *spirituel* that in some contexts might be translated by the seemingly obvious cognates of "spirit" or "spiritual" are, in this particular book, as in other Bachelard works on science, often best rendered as "mind" or "mental."[33] Similarly, the French word *expérience*, which can mean "experience" or "experiment," and on whose ambiguity Bachelard occasionally seems to rely, presents something of a conundrum for the translator. No problem exists, of course, where the context makes the meaning clear. In other instances, where the immediate context is not helpful, I have endeavored to use my best judgment and have taken into consideration the larger context, particularly the overall flow of Bachelard's argument. My aim, here and with the entire translation, is to render in readable English, and as precisely as possible, the meaning of the original French.

What constitutes readable English depends not only on the unique semantic flow and syntactical structure of English but also on chronology. A translation done at the time Bachelard published this book, some eighty-five years ago, would likely differ in some respects from the present one. That is because language changes over time. Thus, in choosing to provide a text in the present-day idiom, to make it consistent with present-day discourse, I have avoided a few outdated practices, including some from even older texts cited by Bachelard. An example would be the occasional use of a capital to begin a term. Although I did keep the capital when the word had an individualized meaning, such as "Relativity" when referring to Einstein's theories, I used the lower case when a more everyday use might be inferred. Similarly, I have followed the usual present-day practice of using the first-person singular instead of the authorial first-person plural when Bachelard speaks directly of his own views or ideas. But, throughout, I have kept in mind the fundamental objective of fidelity to Bachelard's meaning.

ACKNOWLEDGMENTS

IN TRANSLATING A BOOK INCORPORATING PHILOSOPHY AND ITS HISTORY, with particular focus on the history and philosophy of science, I have frequently found it helpful to rely on the knowledge, advice, and assistance of others. I am obliged to Milton S. Feather, professor emeritus of biochemistry at the University of Missouri, and Stephen Danford, professor emeritus of physics and astronomy at the University of North Carolina at Greensboro, for their thorough and careful review of key translated passages on science. My thanks to Adam Graham-Squire, associate professor of mathematics at High Point University, who proved to be a knowledgeable and patient sounding board and source of information on mathematical questions, from geometry to calculus. The consideration of certain philosophical references by Professor Emeritus of Philosophy Terrance McConnell and Dr. Adam Rosenfeld provided essential perspectives for my understanding and my choice of wording, and I thank them both. I am pleased to acknowledge the suggestions and advice of Mary Tiles, whose thorough and highly perceptive work on Bachelard's philosophy of science I have long admired. I am most grateful to Mark Schumacher, senior reference librarian at the University of North Carolina at Greensboro, who was exceptionally helpful in tracking down numerous references and allusions. For her uncommon generosity and discernment in reading and commenting on large portions of my manuscript, I am indebted to Eileen Rizo-Patron, whose contributions to Bachelard scholarship are widely respected and appreciated. I tip my hat to my sons: Roch, Jr., for taking time from his own writing to read portions of the manuscript; Paul, for his discussion of musical theory as it relates to numbers and atomism; and Mark, for his frequent thoughtful considerations of pertinent intellectual issues. Whatever virtues this translation may have derived from the counsel and information I was fortunate to receive from all persons mentioned here, its flaws are decidedly my own. Finally, I wish to thank Elaine, my life's companion, without whose encouragement and support the work on this translation would have been neither attempted nor completed.

—*Roch C. Smith*

NOTES

1. For a complete bibliography of Bachelard's works, as well as an annotated bibliography of secondary sources, see Roch C. Smith, *Gaston Bachelard: Philosopher of Science and Imagination* (revised and updated) (Albany: State University of New York Press, 2016), 153–155.

2. Thomas Kuhn, *The Structure of Scientific Revolutions* (Chicago: University of Chicago Press, 1962), hereafter cited as *Structure*; Gaston Bachelard, *Le nouvel esprit scientifique* (Paris: Alcan, 1934) (*The New Scientific Spirit*, trans. Arthur Goldhammer [Boston: Beacon Press, 1984]).

3. For a discussion of "anomaly," see Kuhn, *Structure*, 52–65. For "epistemological obstacle," see Gaston Bachelard, *The Formation of the Scientific Mind: A Contribution to a Psychoanalysis of Objective Knowledge*, trans. Mary McAllester Jones (Manchester, UK: Clinamen Press, 2002), hereafter cited as *Formation*. The term "epistemological obstacle" is introduced in chapter 1, 24–32, but the concept underlies the entire book.

4. Kuhn, *Structure*, 140.

5. Bachelard, *Formation*, 49.

6. Cristina Chimisso, *Gaston Bachelard: Critic of Science and the Imagination* (London and New York: Routledge, 2001), 73; hereafter cited as *Gaston Bachelard*.

7. Kuhn, *Structure*, v–vi.

8. In addition to Chimisso's 2001 book and my own updated work, both previously cited, see, Miles Kennedy, *Home: A Bachelardian Concrete Metaphysics* (Bern: Peter Lang, 2011); Zbigniew Kotowicz, *Gaston Bachelard: A Philosophy of the Surreal* (Edinburgh: Edinburgh University Press, 2016), hereafter cited as *Philosophy of the Surreal*; and Eileen Rizo-Patron, Edward S. Casey, and Jason Wirth, eds., *Adventures in Phenomenology: Gaston Bachelard* (Albany: State University of New York Press, 2017).

9. Kotowicz, *Philosophy of the Surreal*, 157.

10. These include *Le rationalisme appliqué* (Applied Rationalism) (Paris: Vrin, 1949), *L'activité rationaliste de la physique contemporaine* (The Rationalist Activity of Contemporary Physics) (Paris: Vrin, 1951), and *Le matérialisme rationnel* (Rational Materialism) (Paris: Vrin, 1953). See Smith, *Philosopher of Science and Imagination*, 44.

11. Kotowicz, *Philosophy of the Surreal*, 175.

12. Ibid., 180.

13. Roch C. Smith, *Gaston Bachelard* (Boston: G. K. Hall, 1982), reissued in 2016 in a revised and updated edition; see note 1 in this introduction.

14. Mary Tiles, *Bachelard: Science and Objectivity* (Cambridge: Cambridge University Press, 1984).

15. E. Mary McAllester, ed., *The Philosophy and Poetics of Gaston Bachelard* (Washington: The Center for Advanced Research in Phenomenology & University Press of

America, 1989); Mary McAllester Jones, *Gaston Bachelard, Subversive Humanist: Texts and Readings* (Madison: University of Wisconsin Press, 1991).

16. See Gaston Bachelard, *The Dialectics of Duration*, trans. Mary McAllester Jones (Manchester, UK: Clinamen Press, 2000) (translation of the 1936 *La dialectique de la durée*), and Gaston Bachelard, *Intuition of the Instant*, trans. Eileen Rizo-Patron (Evanston: Northwestern University Press, 2013) (translation of the 1932 *L'intuition de l'instant*).

17. Bachelard, *Formation*.

18. See Smith, *Philosopher of Science and Imagination*, especially 124–126, and 133. See also Roch C. Smith, "Gaston Bachelard," in *The Encyclopedia of Aesthetics*, ed. Michael Kelly (London and New York: Oxford University Press, 1998), 1: 191–195; second edition (2014), 1: 267–271.

19. A discussion of the role of intuition in philosophy would take us far beyond the scope of this brief introduction. I would merely point out that the role of intuition as a means of direct knowledge is ultimately modified with the revolutionary transformation of the atom in contemporary science. As will be seen in the chapters that follow, direct knowledge of phenomena will eventually be replaced by mathematically constructed knowledge and mathematical intuitions will be at work in a science that has become axiomatic.

20. See, especially, *The Formation of the Scientific Mind*, where the word *formation* has pedagogical overtones related to the English word "formative." For a focused and thorough discussion of Bachelard's interest in pedagogical issues within the context of the French system of Bachelard's day, see Chimisso, *Gaston Bachelard*, 51–106.

21. The complementarity of rationality and empiricism will be reflected in the titles of two of Bachelard's later epistemological works: *Le rationalisme appliqué* and *Le matérialisme rationnel*; see note 10 in this introduction.

22. Gaston Bachelard, *La valeur inductive de la relativité* (The Inductive Quality of Relativity) (Paris: Vrin: 1929), hereafter cited as *Relativité*.

23. Gaston Bachelard, *Le pluralisme cohérent de la chimie moderne* (The Coherent Pluralism of Modern Chemistry) (Paris: Vrin, 1932), hereafter cited as *Le pluralisme cohérent*. For a fuller discussion of *La valeur inductive de la relativité* and *Le pluralisme cohérent de la chimie moderne*, see Smith, *Philosopher of Science and Imagination*, 15–21.

24. Bachelard, *Relativité*, 44. The idea of "rupture" or "epistemological break" is central to Bachelard's understanding of scientific revolutions.

25. Ibid., 125.

26. Ibid., 230. (Here Bachelard follows the thought of Arthur Eddington in *Space, Time and Gravitation: An Outline of the General Relativity Theory*, Cambridge: Cambridge University Press, 1920.)

27. Ibid., 155.

28. Bachelard, *Le pluralisme cohérent*, 68.

29. Ibid., 168.

30. Ibid., 182.

31. Ibid., 221.

32. Gaston Bachelard, "Noumène et microphysique," *Études* (Paris: Vrin, 1970), 11–24, originally published in *Recherches philosophiques* (1931–1932): 55–65. Translated by Bernard Roy as "Noumenon and Microphysics," *Philosophical Forum* (2006): 75–84.

33. See, for example, 82, where the French *la connaissance et l'esprit* is translated as "knowledge and mind" and *l'expérience est vraiment une action spirituelle* is rendered as "experience is really a mental action." The titles of two Bachelard books on science are an indication of this dilemma: *Le nouvel esprit scientifique* was published as *The New Scientific Spirit*, while the translation of *La formation de l'esprit scientifique* is *The Formation of the Scientific Mind*. In my view, both are defensible titles, although I would favor the latter as a more appropriate rendition.

PREFACE TO THE FRENCH EDITION

BACHELARD'S SIXTH BOOK, *LES INTUITIONS ATOMISTIQUES* WAS PUBLISHED for the first time in Paris, in 1933, by Boivin. The work followed *Le pluralisme cohérent de la chimie moderne* (The Coherent Pluralism of Modern Chemistry) (1932) and *L'intuition de l'instant* (1932).[1]

The philosophical notion of "intuition" (from the Latin *intueri*, to perceive), used in various ways by Descartes, Kant, or Bergson, always retained an immutable character for these authors. But in *Intuition of the Instant* Bachelard had shown that the focus of time was modified with advances in physics. *Le pluralisme cohérent de la chimie moderne*, for its part, revealed that contemporary chemistry turned its back on substantialism. The intuition of matter made up of atoms that are solid and indivisible (the very meaning of *atomos*) could thus not be definitive. Multiple in and of itself, not very coherent if not already contradictory in the eyes of historians, the doctrine of Greek atomists, as Tannery,[2] Metzger,[3] or Brunschvicg had already remarked, was barely a philosophy. Rather, it involved a "world of mingled images and reasons" (1) that persisted up to our own time. Bachelard undertakes to inventory these images and to index them. While, for both pedagogical and philosophical reasons (to help clarify ideas and compare doctrines), it may initially be a matter of finding the "fixed points" in a sort of triangulation of doctrines (2), Bachelard does not exclude the possibility that this approach may also play a cathartic role with respect to science[4] (9). To get it started, he makes use of a method inspired by his work on chemistry, consisting of three components:

1) As Lavoisier or Mendeleev had done for chemical compounds, our philosopher sets out here to reduce doctrines of the atom to simple elements (intuitions and/ or arguments) that can then be sorted out. The resulting classification will bring together the various kinds of inventoried atomisms (realist, positivist, critical, and axiomatic), the last three being forms of idealism.

2)   While insisting on the changes taking place, the method seeks to reveal, first,
      how these forms of atomism conceive of synthesis or composition.[5] For the
      atomism of antiquity, if a rational (pseudoscientific) view seems to prevail, it is
      especially because emphasis is placed on the theoretical side of knowledge, to
      the detriment of the experimental side (4). Hence, the atoms of Democritus have
      perfect properties (hardness, immutability, permanence, etc.) and hypothetical
      hooks for attaching, while those of Lucretius, subject to the *clinamen*,[6] are sup-
      posed to explain freedom. But who does not see how far we are from real science?
      "Nowadays," writes Bachelard, "scholars who refuse to associate the philosophies
      of Democritus and Lucretius with modern scientific atomism are numerous" (6).
      He himself will not grant that they influenced the atomists of the Classical Age
      (Gassendi, Huygens, Boyle) or of the early nineteenth century (Dalton).[7]

3)   Bachelard thus avoids Brunschvicg's continuity, but, even so, he does not conclude
      that knowledge is totally discontinuous. Of course the philosophers of antiquity
      are not precursors: in the history of atomism "there is nothing similar to those
      influences that span the centuries," and Democritus is not "the first proponent" of
      positivist experimental thought (7). Nevertheless, our philosopher observes that,
      in atomism, the appeal to experience is steadfast, and that atomistic philosophy, in
      general, "enjoys such a clear dialectic that, in every period, the same duality and
      the same divisions between the various ways of conceiving of the atom reappear"
      (7). From precisely this stems the possibility of a classification that presumes a
      certain license with history: he will feel free to dismantle systems,[8] to mingle time
      periods, and occasionally to drop the incidental or the "specifically historical" (8).

Starting with the liberating experience of dust[9] (15), another indicator of the
void that allows an early refutation of the link Bergson sees between intelligence and solidity (17), atomistic thought will need to find a way to break from
the various forms of realism that hamper it and, from Democritus to Lémery
and beyond, to refine its theses. It will even be necessary for the atom to take
on a punctiform aspect (Boscovich) (41), to lose its main characteristic (indivisibility) and become severable before we can finally get to modern physics.
With positivist atomism (Kirchberger, Proust, Berthelot) and its insistence
on experimental verification (influenced by Richter, Dalton, and Lange),
"materialism" will be able to grow stronger and the chemical phenomenon,
with Avogadro, then Perrin, will itself end up submitting to the atomistic
hypothesis. Beyond the neo-Kantian criticism of Hannequin (67 ff.) we will
then enter, at the beginning of the twentieth century, the period of principles
and postulates, along with their mathematical coordination, namely, axiomatic
atomism (85 ff.).

From his "taxonomy" Bachelard derives major philosophical conclusions: 1) the order of ideas energizes ideas, and it is through such ordering and composition of ideas more than through their analysis that thought can lead to discoveries (87). 2) Over the years, atomistics becomes an instrumental atomistics that produces (not finds) precise, schematic phenomena, "immersed in theory"[10] (89). The instrument itself, according to a still renowned formulation, is but a "reified theorem" (90). 3) Modern science, thereby, tends more and more to become a science of effects (Zeeman, Stark, Compton, Raman effects . . .) that are mathematically described and closely linked. 4) The electron's complex character, which eludes the laws of classical electrodynamics (92)—and not its simple character, as Meyerson believed—forces Bohr to invent a new physics, a "non-Maxwellian physics" that counters common sense and in which the classical intuitions of trajectory and space (95) no longer apply.

In conclusion, scientific investigations into atomic phenomena, according to Bachelard, underscore the deceptive character of our first intuitions that "respond too soon and too completely to the questions that are posed" (97); they also prompt us to reject a realism that "attributes to the scientific object more properties than we actually know about it" (97); and finally, in order to be correctly interpreted, they commit us to associating them with multiple philosophical perspectives[11]—criticism, positivism, idealism, and even realism (as long as the latter is limited to a simple inclination as a function of successive enhancements of knowledge)—all of which need to be put in order (98).

Thus, without yet plainly entering the new physics,[12] Bachelard has nonetheless captured its essence: an experience enriched by a considerable range of mathematical assumptions, but one that remains of profound interest to the philosopher, for it carries within itself "the most prodigious of metaphysics."[13] As for the "classificatory" method used by the author, it is still today at the heart of numerous investigations.[14]

—*Daniel Parrochia*
September 2015

NOTES

1. [Translated by Eileen Rizo-Patron as *Intuition of the Instant* (Evanston: Northwestern University Press, 2013). Titles of published translations will be used hereafter.]

2. P[aul] Tannery, "Qu'est-ce que l'atomisme?," in *Mémoires scientifiques*, vol. 8, *Philosophie moderne, 1876–1903* (Toulouse: Privat, 1927), 303–337.

3. H[élène] Metzger, "Compte-rendu des *Intuitions atomistiques*," *Revue philosophique de la France et de l'etranger* 116 (July–December 1933), 310–312.

4. His book *The Formation of the Scientific Mind* [1938] will take up this project.

5. Similarly, later on, in an approach that is also inspired by scientific taxonomy, Bachelard will reduce the complexity of the forms of imagination to the four pre-Socratic elements (water, air, earth, fire) in order to explain their combination.

6. [See Introduction, 103, note 5.]

7. This assessment has been contested. See J. Salem, ed., *L'atomisme au XVIIe siècle et XVIIIe siècles* (Paris: Publications de la Sorbonne, 1999), preface, 7.

8. Bachelard will return to this idea in *The Philosophy of No* [1940].

9. F[rançois] Dagognet will praise this as well. See F[rançois] Dagognet, *Pour le moins* (Paris: Les Belles Lettres, 2009), 29–46.

10. A closely related expression (*theory laden* in English) can be found in Quine. See W. V. Quine, *Pursuit of Truth*, revised edition (Cambridge, MA; London, England: Harvard University Press), 7; [The editor refers to the French translation of Quine by Maurice Clavelin, *La poursuite de la vérité* (Paris: Seuil, 1993), 28.]

11. Such a polyphilosophy will find its theory, here again, in *The Philosophy of No*.

12. It may surprise us that this book is virtually silent on the works of the quantum physicists of the time. There are, to be sure, a few references to Millikan, Stern-Gerlach, or Broglie. But the approach to Bohr is not very technical and Heisenberg is only cited once. This can be explained by the dates: in the 1930s, quantum physics was still recent and most of the elementary particles remained to be discovered. We will have to wait for Bachelard's later books, such as *L'expérience de l'espace dans la physique contemporaine* (The Experience of Space in Contemporary Physics) or *L'activité rationaliste de la physique contemporaine* (The Rationalist Activity of Contemporary Physics), for more precise analyses of the rich reality of the world of particles.

13. One that is henceforth explicit. See, for example, S[ven] Ortoli, J[ean]-P[ierre] Pharabod, *Métaphysique quantique* (Paris: La Découverte, 2011).

14. See F[rançois] Dagognet, *Tableaux et langages de la chimie* (Paris: Seuil, 1969); *Le catalogue de la vie* (Paris: Presses Universitaires de France, 1970). See also D[aniel] Parrochia, P[ierre] Neuville, *Taxinomie et réalité, vers une classification* (London: Iste, 2014).

INTRODUCTION

The Fundamental Complexity of Atomistics

I

IT IS THE MISFORTUNE OF ALL GRAND DOCTRINES TO ENTER INTO CONTRA-
diction as they evolve and to be unable to flourish without losing their original
purity and clarity. The definitions at their base grow obscure with repeated
application. The words themselves abandon their roots as usage tarnishes
their etymologies. The well-chosen convention to which these words initially
pointed soon becomes a mere rule. In other words, a limited meaning, precise
enough to clarify a truly useful idea, calls forth a wider meaning through its
very use. The fact that an idea comes to contradict etymologically the very
term that represents it by extending its reach in this way does not, in and of
itself, constitute a decisive objection to such a notion. Rather, it would be a
sign that the idea has left the world of simple definitions to become a veritable categorem.[1]

Léon Brunschvicg[2] shows that, early on, from Democritus to Lucretius,
a contradiction took hold within the atomic hypothesis, and that two great
doctrines, brought together under the same sign, but with different goals and
destinies, moved forward together until the scientific era. Thus, atomism seems
to have assimilated its opposite from the very first attempt to expand it. Very
quickly it passed from a realistic meaning to a categorematic one. The atom,
taken initially as an object of intuition, furnished an opportunity to think in
terms of a method for discursive analysis of the phenomenon. A whole world of
mingled images and reasons was thus already latent within the first doctrines
of atomism. This mingled form would naturally persist when philosophical
developments began to enrich the doctrines.

Under these conditions, it may be best to begin with an analysis, and
even a dismantling, in order to isolate the disparate elements of doctrines

that hide such varied thoughts under the same name. My goal has been to prepare this analysis and to furnish students with the means and pretexts to classify their ideas. Arguably, my work with individual systems that makes possible an understanding of the whole is not likely to be a distraction to this group. If my analyses have meaning, they will do no more than facilitate the comprehension and especially the comparison of the doctrines. A few clearly detached elements can, in effect, serve as a point of focus. All triangulation requires fixed and clearly visible points. If the elements that I isolate correspond to salient facts, the triangulation I propose can furnish a map for the detailed description of the systems.

Let us start right away with a feature that can help us bring together the scattered chapters of this little book. This feature will show that I myself would hesitate to place in definite opposition the doctrines I have distinguished. It seems to me, in fact, that the two directions identified by Brunschvicg's initial explanation of the atom are so exactly inverse that, more than analytical paths, they indicate a back-and-forth epistemological movement that is equally clear and productive. In other words, the antisymmetry of the doctrines is so perfect that it reveals a certain solidarity in the solutions rather than a heterogeneity of the objects under investigation. In fact, two systems of thought uncovering the same elements, in the same relation, in the same general order, only in opposite directions, are basically reducible to a single form. These two systems, in short, follow the parallel but inverse movements of analysis and synthesis. Rather than being opposed, they are complementary. They are verified one by the other and it would be vain to attempt to destroy their solidarity, to include one by excluding the other.

In the atomistic world, analysis and synthesis have such a precise, material, and general meaning that it may be good to insist on the rhythm of reciprocal proof that these two types of thought, explication, and experiment take with regard to each other.

As one of its main ideas this book will show that it is really the atom that is sought when the phenomenon is analyzed, but that, at the same time, atomism is justified only through synthesis, by indicating how we can develop a *composition*. Proof by means of an ultimate element benefiting from an evident reality, by an atom held in our fingertips as a result of analysis and answering all questions by its mere reality, would be definitive. This would be a sort of absolute analysis that escaped from reciprocity. Such a method would finally replace "why?" with "how?" And yet one question would have been left out, a last refuge of an insurmountable "why?": in effect, who will explain *composition*? In thinking over the problem, we notice that reasoning that involves the simple composition of two atoms cannot reside entirely in the *nature* of each of the two atoms. Thus, we face two conclusions that are equally necessary yet divergent. On the one hand, if the component element could accommodate all the characteristics of the compound, we would have to conclude that, in reality, there is no composition. And so an explanation that starts off from too substantial an atom is entirely verbal. On the other hand, it is quite certain that the loosest and simplest compositions, such as juxtaposition or mixing, for example, derive at least some of their explanatory power from space. It can be seen in this case that the atom is not self-sufficient, that an *outside* must necessarily be attributed to it, and that relations with the exterior constitute a kind of second-order reality that sooner or later enriches atoms that were once posited as extremely meager. Thus, as many examples will show, either the atom is too rich and the problem of composition—albeit a real one—has no meaning, or the atom is impoverished and composition is incomprehensible.

Hence it is useless to seek an absolute analysis. We will always have to judge an analysis by the synthesis it favors. Similarly, a synthesis will only be understood thanks to a preceding analysis. It is by joining analysis and synthesis that we recognize the full worth of these two modes of thought.

If, therefore, in dealing with a specific problem, we chance upon a reciprocity of movement that is as exactly complementary as the one observed by Brunschvicg at the center of the atomistic account, we have some assurance of possessing a valid explanatory rhythm, on condition of uniting both features. We have an association of thoughts that is at once correct and objective. The *object* is not in one direction over another, or, to put it differently, objectification will not occur by analysis or by synthesis alone, for objectification is produced by the correct and clear pairing of analysis and synthesis. That account's perfect reversibility reconciles the logical and empirical qualities of knowledge. It represents a maximum of homogeneity at the heart of experimental knowledge.

Of course, little of this homogeneity is found in the doctrines of antiquity, and it is quite certain that Brunschvicg could note, between Democritus and Lucretius, the contrast evident in simple thoughts from the moment they differed. Keeping in mind our prior reservations, these two initial forms of atomism can thus serve as indicators that will classify, right from the beginning, the features of our problem. I shall characterize a little more closely these two *epistemological directions*.

II

TO BEGIN WITH, WHAT DIRECTION DOES THE DEMOCRITEAN ACCOUNT TAKE?
And, first of all, what is its point of departure?

In this doctrine one starts by breaking outright with the qualities of the
phenomenon. Entirely incongruous and even opposite characteristics from
those apparent in the phenomenon are attributed to the elementary corpuscles
that will determine the whole explanation. In this way the atom will be given
perfect properties: hardness, immutability, permanence, disposition toward
geometric form and symmetry. In essence, initial atomistic thought thus seems
a truly audacious theory. It does not hesitate to diverge from experience in
order to impose a *rational* view of reality.

It has often been said that the Democritean school was inspired by a
true scientific spirit. Yet that is not enough to characterize this school, for the
scientific spirit is twofold at the very least, depending on whether it accentuates the theoretical or the experimental side of knowledge. The early Greek
atomists seem to me to be headed in the first direction, although they are not
aware of it. They believe they are observing, but they are already reasoning.
Also, my overall view is compatible with the historical judgment of Bréhier,[3]
who recalls the life of travels and observations of Leucippus and Democritus.
Henceforth, in following the fate of the Democritean intuition all the way
to modern thought we will necessarily face a clearly and economically constructed atomism. We will see a veritable axiomatics of the atom develop
along these lines. In other words, this path to understanding atomism will be
revealed, in certain respects, to be nothing other than the *corpus* of postulates
that are indispensable to a geometric and mechanical explanation of the phenomenon. As a result, I will be able to say, in one of my conclusions, that the
atom embodies the sufficient, if not the necessary conditions for a theoretical
construction of the phenomenon.

Of course, the point of view attributed to followers of Democritus is
not as neatly unified as my extreme schematization of frequently mixed perspectives would indicate. I'm not unaware, in particular, that we are usually
justified in recalling the experimental character of their epistemology, especially when it is set off against the metaphysics of opposing schools. But, in
my view, the experimental portion of their doctrine, seen from a rationalistic
perspective, is weak because it seems entirely incongruous with the body of
the general commentary. To the extent it draws its inspiration from the phenomenon, the structure is poorly adapted to the atomistic characteristics

that have been postulated. This structure seeks to recover the phenomenon without following a truly mathematical progression. Had it developed along purely logical lines, following the value given by a rational combination of postulated elements, it might have missed out on a synthesis from an experimental point of view, but, at least, it would have been an intrinsically correct synthesis. Moreover, the circumstances of such a failure might have led to a rectification of the point of departure. On the contrary, since a latent pragmatism constantly distorts logical development, we do not see the conditions of a healthy verification show up in the physical science of antiquity. In the end, an analysis that claims to specify the characteristics of the atom and a synthesis that claims to construct the phenomenon are disjointed. They do not connect, so they cannot verify each other. One might as well say that the experimental and the theoretical efforts of the doctrine obey two uncongenial impulses and that, with Democritus, the scientific mind has not yet been able to draw together two currents that find in their convergence the unity of the phenomenon and rational certitude.

LET US NOW TRY TO IDENTIFY, IN THE EPICUREAN ACCOUNT, THE FEATURE that can provide a new indicator for the classification of an entire category of atomistic doctrines.

With this dominant feature, Epicurean thought, far from breaking with common experience from the outset, willingly takes ready-made properties from the overall phenomenon and carries them over to the explanatory element. To be sure, as I have just pointed out, Democritus, like all positivists, was not able to exorcise the explanation's finality; but, at least, he made a great effort to hide it and to reduce it. While in fact guided by phenomenal characteristics, his system claims to construct them. With Lucretius, on the other hand, the phenomenal characteristic is clarified at the level of explicatory postulates themselves. Brunschvicg provides a demonstration of this in a special case. Freedom is surely what is most difficult to construct.[4] Since Democritean developments do not accomplish it, we find ourselves, within strict Democritean doctrine, affirming a kind of determinism. We should point out that such a determinism is put forward as a hypothesis. No experiment proves it or even points to it. Epicurean doctrines, on the other hand, accord a veritable freedom to atoms with the assumption of uncaused deviation, of the *clinamen*[5] that requires no explanation since it is attributed directly to the atom. Thus, the atom encloses within itself all the exterior properties of freedom. One can appreciate how easy it becomes, in a world with this kind of relaxed determinism, to insert human freedom with all its characteristics, its development, and its various impulses. But such a deduction immediately has the makings of a vicious circle since we are limited to rediscovering what had been postulated.

And so, on the specific question of the role and the place of freedom in the synthesis of the phenomenon, an opposition can be seen between the two kinds of doctrines that stem from Democritus and Lucretius. In one system, the solution is impossible; in the other, it is, so to speak, too easy. To characterize this opposition by going back to the very essence of the general methods alluded to earlier, we can say that, in the doctrines inspired by Democritus, there is a failure of synthesis. On the other hand, in the doctrines stemming from Lucretius, there is no real in-depth epistemological movement, no real analysis. In both cases, we are far from having associated analysis and synthesis with a view to mutual verification, since we clearly remain lodged within the framework of the initial hypothesis.

Finally, another conclusion follows upon this initial rough assessment, which is that the thought of Democritus, while the most learned, seems to borrow the fewest elements from reality. It will always be associated with an idealist philosophy. By contrast, the thought of Lucretius, less strict and less careful in its choice of bases, seems to be closer to the phenomenon and ultimately more realist.

III

THUS, PERHAPS I WAS RIGHT TO ASSERT THAT ONE OF THE SYSTEMS DOES NOT continue the other and that, after Lucretius, atomism is revisited and rethought from its very foundation and for entirely new purposes. This power of originality and renewal, easily masked by identical terminology, persists in fact in more recent atomistic schools. If my goal were to retrace the historical development of atomistic doctrines—really an unnecessary task after Lasswitz's admirable work[6]—I would find myself frequently called upon to point out the same disparity of methods and the same fragmentation of conclusions. There are perhaps few clearer examples in philosophy of the independence and isolation of doctrines than in the development of atomism. Nowadays scholars who refuse to associate the philosophies of Democritus and Lucretius with modern scientific atomism are numerous. I would venture to go further. The atomistic doctrines of antiquity do not seem to me to have had any real influence in modern times. They did not really inspire the theories of Gassendi, Huygens, and Boyle, nor

Dalton's research. In fact, the basically immediate intuition that gives each of us the fundamental traits of the atomic model cannot be considered real learning. For atomism there is nothing similar to those influences that span the centuries and that, at times muffled, at times conspicuous, carry Platonism, Cartesianism, and pantheism to the very heart of the most varied doctrines, enrich thought, and correlate systems. For example, when Bacon cites Democritus, it is really only to credit his use of the *word* atom. At most, he recognizes in the Greek philosopher the master of a declared and methodical aversion to metaphysics. That should not be enough to suggest that Democritus is the first proponent of experimental and positivist thought. Nevertheless, this opposition to metaphysical thought—however obscure and even inexact it may appear when examined a little more closely—amounts to referring atomism to experience alone. And such recourse to experience, which can give the doctrine a guarantee of permanence, also explains the spread of this doctrine without our even having to speak of influence from thinker to thinker.

In fact, once intuition has taken experience as its point of departure, it can develop further by yielding to the actual power of experience. If, moreover, we add that it *must* develop in this way, namely, that the first task must be to put aside learned suggestions and look at facts with fresh eyes, we will understand that atomism is almost always presented in the history of philosophy as a reaction to history, as a declaration of the right to treat the problem of the real through direct experience.

However, these claims to being scientific fall short, and centuries go by before they can form a general method. Moreover, the metaphysical mind does not relinquish atomistic doctrines through mere statement, and when it comes to the very specific concept of the atom, the most varied ideas—including the most personal—join in clearly arbitrary constructions. Is there a more mixed body of doctrines than atomism taken as a whole? Does it not go from materialism to monadism? From material unity, with a monist quality that is barely distinguished by spatial characteristics, to the most profligate phenomenal diversity? How can we resolve the apparent contradiction between the simplicity and uniformity of the point of departure and the complexity of developments? It may suffice to point out that, on the one hand, what is transmitted is a word and an invitation to experience—a reason for stability and conformity—and that, on the other hand, what unfolds is a philosophy like the others where individual intuition is marked by its own fancy.

As a matter of fact, this atomistic philosophy enjoys such a clear dialectic that, in every period, the same duality and the same divisions among the various ways of conceiving the atom reappear without much variation.

Renouvier pointed out that the pre-Socratic philosophies were divided "into as many doctrines as it may be possible to posit general principles and their opposites to explain the nature and cause of beings."[7]

That is even truer of atomistic doctrines. We can thus hope to find a clear if not rational classification, despite the historical diversity of such doctrines.

IV

SUCH AN OBSERVATION PARTLY JUSTIFIES PERHAPS THE EXPOSITORY METHOD I have chosen in these inquiries. As I have said previously, my goal is to underscore the intuitive traits of atomistic doctrines, to show also how an intuition becomes an argument, and how, finally, an argument seeks out an intuition to become clearer. I found it necessary to dismantle the systems in order to separate out their elements distinctly. Under these conditions, I reserve the right to borrow examples from quite different moments of philosophical development. I shall shuffle periods rather than genres. I shall also discard what is incidental and specifically historical in certain conceptions. The history of philosophy being a history of reason and experiment, it may be useful to delineate the basic principles of a reason and an experiment from time to time. If I thus succeed in identifying some of these essential principles of atomistic philosophy, while providing an initial, provisional classification of its several intuitions and arguments, the reader of this book may then be able to read fuller books more rapidly and compare with greater clarity the innumerable works of atomistic philosophers. It is toward this simple task, a quite preliminary and pedagogic one, that I hope ultimately to have worked.

HERE THEN, IN BROAD OUTLINE, IS THE PROGRAM FOR THESE INQUIRIES. Following the very path of duality that I identified by way of introduction, I have divided my investigations into two series of chapters.

I will start with *atomism related to the realist schools*—the simplest, most naïve of atomisms—endeavoring to show how it fits into a wider realism. However, in order to tackle its examination more readily, I will begin by what I consider to be the intuitive basis of all atomism. Once the means of knowing or the occasions for imagining have been isolated, we will be better placed to appreciate the scope of metaphysical thinking. Then it will be clearer that realist atomism is a metaphysics like any other, which is to say, remote from experimental verification.

Before moving on to the other schools, I will show that realist atomism dismisses an essential problem that needs clarification: the question of phenomenal composition. I will devote a short chapter to it.

In the second part of my work I will then examine, always in the same spirit of free and factitious analysis, various types of atomism that are more or less closely associated with idealist philosophy.[8] I will make the following distinctions:

> *Positivist atomism*, so skillful and so excessive in its restrictions that it sometimes finds a way to pass as realist in its experimental affirmations, all the while being incontestably idealist with respect to the hypothesis that holds it all up;

> *Critical atomism*, able to associate with the most varied scientific theses;

And finally I will address the principles of modern scientific atomism. Without going into properly scientific territory, I will identify several philosophical principles that mark modern atomistic thought with brand new traits. It is here that we will see efforts of reason and experiment converge. It will then be a question of recognizing the logic of experimental research, of gathering axioms, of preparing theorems, and of producing the physical *effects* anticipated by mathematical physics. The role and the place of intuitions will be turned upside down. Intuitions will no longer be *established particulars* to be developed and organized, but simply *figures* that give voice to what we say. Modern atomism will come across as essentially discursive, and will take great care to avoid a priori metaphysical intuitions. It will replace initial images with axioms, or, rather, it will accept such images only as figures used to illustrate axioms. In the area of our present inquiry, the *systematics of assumptions* that characterize modern science might be thought to give legitimacy to the term I am proposing of *axiomatic atomism*.

Thus, if my work, in general, is to have meaning for the study of the principles of contemporary science, we should see it as having a cathartic function. It is by knowing traditional metaphysical intuitions in a discursive and detailed manner that we will be more easily able to put a stop to the exaggerated action of these intuitions in an area where they can no longer be any more than metaphors. Faced with matter that is infinitely small, we confront a break in our experience. In order to examine it, reason must be allowed free rein. In other words, contemporary microphysics is the science of a new world, and a "metamicrophysics" must be grounded on new experiments with new categories.

PART ONE

Chapter I

THE METAPHYSICS OF DUST

I

IF EVERYDAY EXPERIENCE DID NOT PROVIDE US WITH THE MANY AND VARIED phenomena of dust, it can be assumed that atomism might not have received such a ready following from philosophers and that it might not have enjoyed such an easily renewed fate. Without this special experience, atomism could only have been conceived as a highly speculative scholarly doctrine in which the idea's initial venture was not justified by any observation.

By contrast, based solely on the existence of dust, atomism has been able to receive, from the very beginning, an intuitive base at once permanent and rich in suggestions. These initial suggestions evidently serve to explain atomism's historical as well as its pedagogical success and, here in particular, philosophy benefits from bringing together pedagogical and historical elements. From this straightforward pedagogical perspective, I will try, in a few pages, to study atomism's simplest image, one that is durable precisely because it is simple and rudimentary. Charles Adam, for example, did not hesitate to see Descartes's younger days as the source of some of his guiding intuitions. As he points out, because Descartes lived in the country, he was able to take note of several curious traits of nature. Among such natural lessons, Charles Adam specifically includes familiarity with phenomena like will-o'-the-wisp, dust, and whirlwinds.[1] In fact, it should be noted that a *whirlwind* is a rarer occurrence than one might think and that many talk about it who have not had the opportunity to observe it. One must have seen the dust on the road, at the bottom of a ravine, caught up and lifted by a favorable wind to understand what is at once structured and free, light and delicate, in the swirls of a whirlwind. The best-made whirlwinds are the smallest ones. They stay within a wheel path. They can actually rotate around themselves like a humming top. More commonly observed river eddies give us a far cruder image than a whirlwind drawn by dust. Water only gives us a lightly engraved design; dust gives it in full three-dimensional relief.

Whatever one may think of the importance attributed by Charles Adam to these first material images of Cartesianism, there is no doubt that one finds, in what is most often a radically materialist atomistic literature, numerous quotations pertaining to phenomena of dust. It therefore seems astonishing that Lasswitz does not include in his otherwise detailed index anything that recalls ideas of dust, powder, or pulverization.[2] These concepts certainly deserve to be given priority over amber, mercury, and smoke—which Lasswitz did include.

II

FOLLOWING THESE GENERAL REMARKS, LET US ATTEMPT TO APPRECIATE the importance of dust for the teaching of atomism.

We can start by presenting something of a negative proof of the intuitive value of such a phenomenon. All it takes is to imagine how our intuition would be affected by a world of well-defined solids, a world of objects whose individuality would be strongly and clearly related to size, as is the case, for example, for all animated bodies. For greater clarity, let us complete our assumptions by setting up a world where these defined and individualized objects have sizes that extend over a rather limited range, thus containing neither very large nor very small objects. We understand right away that in such a world material division would be seen solely as an *artificial* process. Intuitively speaking, we could shatter, but we could not analyze. Of course, an advanced science might succeed in transferring the principle of individuality elsewhere, agreeing, for example to analyze a solid *geometrically*. But then geometrical analysis and the partition of the real would no longer be synchronous. The former, bearing the mark of ideality, would belong to the world of possibility pure and simple. Nothing real would correspond to it.

Now let us change scientific utopias. Instead of a world of well-defined geometric solids, let us imagine a world of pasty objects, such as, for example, a universe briefly considered by Mach,[3] that is a little too hot, where everything flattens out, where forms inhabited by essential fluidity are nothing more than moments of development. This time, contrary to what happens in our first hypothesis, division is now the law. Every object dissolves, loses its shape, and is endlessly segmented. The ideal pattern is flowing water that divides as easily as it reassembles, thus illuminating a perfect reciprocity of analysis and synthesis. Faced with such a scene, how could we posit the idea of an *indivisible element*? The only way would be to contradict concrete experience and generally observable evidence. And here again our means of separating the real and the possible would be deeply perturbed. Yet all we did was put forth a poor, simple assumption as we constituted our scientific utopia, only to see that assumption modify all that is possible and, like a reagent, precipitate a brand-new reality! In a world of pastes and liquids, it would seem that the possible is, I daresay, more real than immediate reality. For the possible is now everything in the process of becoming, now rendered more clearly by its increased activity. By contrast, reality is nothing more than an ephemeral and accidental form, an individual frame in a film. By underscoring through thought the fluidity of solid bodies, we might have believed that we affected only a material quality, but we realize, in the end, that we have perturbed even the most fundamental categories and forms of our knowledge since we enter into an extraordinary world where time finally dominates space.

So, in a way, we can frame the real world with two hypothetical worlds that are equally easy to imagine: the first where solidity is everything, the second where solidity is nothing. But one can see right away that, in these two utopian worlds, atomism does not encounter the elements of its first teachings since, in one of the hypotheses, division of matter would be an anomaly and, in the other, a rule to be endlessly applied. Realistic atomism is indeed dependent on a direct intuition of material diversity. I have tried to show elsewhere how difficult it is for scientific thought to uncover categories and order within immediate diversity.[4] In some ways this diversity must be seen as irreducible if we want to preserve atomism's full value of explanation. That is why, as we have just seen, atomism immediately loses all meaning when a profound, hypothetical cause of uniformity is slipped into the real. The concept of dust, halfway between that of a solid and a liquid, will, by contrast, furnish a sufficiently mixed proof on which to base atomism.

Of course, as I indicated earlier, this is but a negative argument, one that tends to underscore atomistic philosophy's dependence on the very general empirical conditions in which thought is developed. I must now begin a more positive examination and take things as they are, not only in their multiple forms but also in their frequent deformations.

III

THE THESIS THAT I WILL DEFEND, AT ONCE GENERAL AND COMPLEX, GOES against Bergsonian theory in that it sets out to complete a proposition that, in its very essence, should not require completion.[5] Indeed, Bergson undertook to assimilate our fundamental habits of thought to our everyday experience of solids. According to him, everything that is framed, categorical, and conceptual in human intelligence stems from the geometric aspects of a world of solids. Our experience of solids leads, in a way, to solidifying our actions. Objectivity as stability is thus related to the solidity of objects. Only what is solid is thought to hold a sufficient number of features strongly enough to represent and maintain the "dotted line" that outlines our possible action. In the face of the simple sketch of our actions thus geometrized through our experience of solidity, all other natural phenomena come across as irrational.

Bergson has surely uncovered in this instance a dominant feature of understanding. In particular, everything that is exchanged socially is expressed in the language of solidity. Similarly, a substantive noun is, in effect, defined from the outside. It can be placed in any sentence the way a solid is placed in any location. In its logical form, language thus corresponds to a geometry of the well-defined solid. But here is where Bergson's thesis needs to be extended. If the initial orientation of intellectual and verbal organization really means the immediate utilization of objects of experience, how do we delete equally characteristic elements from that everyday experience? How do we overlook flowing water, silent oil, sticky honey, paste, mud, clay, powder, and dust? To be sure, all these things find their way back to solidity, but they also contradict certain essential characteristics of solids. Let us not object that solidity is the rule and that liquid or dust are exceptions. For it is quite remarkable that, as principles of explanation, clear and flagrant exceptions carry the same quotient of conviction as do general characteristics—a strange dialectic this, one that thrives on oppositions yet rejects from the bases of its explanation only those elements that are mixed and mingled! Even from a scientific perspective, are the most frequent themes of an explanation not the perfect, undeformed solid and the perfect liquid without viscosity, in other words, two features that are frankly exceptional? One has to arrive at a very advanced physics to find any appeal in the study of mesomorphic states.[6] But from a psychological point of view—the only one that interests me at the moment—these studies of intermediate states are analytical; they are expressed with the help of supposedly simple primary states. At the same time, states taken to be primary—solids, liquids, paste, or dust—do not raise *questions*; they provide the direct *answers* of intuition. They are *elements of naïve explanation*. As a result, it is all of nature that teaches us, and understanding enters through all our senses. Thus, we must speak of a kinetic intelligence alongside the geometric intelligence given primacy by Bergson. We must even add a materialist intelligence. Ultimately, we must recognize that our language is, if not by its nouns at least by its verbs, as tactile as it is visual. Henceforth a more objective intuition of matter will lead to what is, from many points of view, a broader Bergsonism.

In my view, a deformation, even when visual, is not understood as a mere loss of forms, for as soon as we consider how our actions are accomplished, we realize that the deformation we impose on things always means actively acquired information. And so it is a question of taking shape, often with great difficulty, rather than losing shape. Thus, we come to experience deformation as dynamism. For example, the idea of penetrability acquired in the potter's arduous manual experience proves to be fundamental. Henceforth an impenetrable solid is seen as an outright exception. The outline of its shape corresponds to nothing more than our idleness, a prospect of laziness, and a philosophy of the immediate. If we wish to relate Homo sapiens to Homo faber, we must consider the latter in all manner of actions. Homo faber arranges and kneads; such an individual welds and grinds. For that person certain bodies are juxtaposed, others are mixed together, and still others are dispersed in dust and smoke. Solids demonstrate the great lesson of form and assembly. From liquids comes the equally fruitful and clear lesson of change and mixture. From the phenomena of dust, powder, and smoke, Homo faber learns to meditate upon the delicate structure and the mysterious power of the infinitely small; along this path lies the knowledge of the impalpable and the invisible.

And so the primacy of explanation via solids is compromised at the very core of popular knowledge, in the domain of initial intuitions. Besides, even if we were to assume that the problem of the intuitive origin of knowledge remained unresolved, we would at least have to admit that the characteristic of absolute solidity attributed to bodies is a *characteristic to be rectified*, since the best-known phenomena soon display a departure from the quality of perfect solidity. In reality, our thinking is more in line with the deformation of a body than with the geometric relation between many bodies. Thus, Bergson's thesis designates only a point of departure. It is unable to account for the complete evolution of objective thought.

IN SHORT, WHETHER THROUGH UTOPIAN ASSUMPTIONS, OR THROUGH glimpses that describe matter in the actuality of its multiple states, I believe I have restored to my intuition an unfocused and free character brought about by several sensory sources. It will now be easier to sever the link that is always too narrowly established between principles of atomism and geometric intuitions derived from observing solids. Following this polemical preparation, let me now move to a truly positive examination of my thesis. Let me attempt to demonstrate that the intuition of phenomena of dust truly undergirds naïve atomism.

IV

WE SHOULD, FIRST OF ALL, RECOGNIZE AS FACT WHAT IN FACT EXISTS. NOW the experience we have of powder and dust is far from negligible. This experience is so singular and striking that we can speak of a *powdery* state exactly in the same way that we speak of solid, liquid, gaseous, and pasty states. In reality, in modern science this powdery state always poses problems of its own. For example, we see a more energetic chemical action in powders. This chemical potency of powder derives from a kind of surfacing. Zones of transition and contact will give way to special phenomena. Catalytic actions appear that would have no impact coming from a material taken as a mass. Thus, Auguste Lumière points out that exchanges and reactions that take place in the tissues of a human adult extend to a surface of two million square meters: "However minute the affinities may be of substances coming into contact on the periphery of granules, we can conceive that the sum of all these infinitesimal elementary reactions can become considerable when occurring over such large surfaces."[7] We might thus say that, through granulation, *surface* takes on an authentic substantial reality. It ceases to be geometric to become truly chemical.

Even from a coarser and more mechanical perspective, powders work in a special way; their drift and flow lead us to study carefully the shape of their containers or the partitions along which they must slide. But it might be objected that this too is a new and delicate technique. So let us locate the freshest intuition possible.

Let us first consider a child's amused attention before an hourglass. Let us contemplate, along with that child, a complex of exceptions! Powder is solid, yet it flows; it falls noiselessly. Overall, surfaces are at once mobile and stable. Mounds will grow; craters will form in which one can see uncaused movement begin. If now we try to reconstruct the overall phenomenon starting from the movement of separate particles, we are amazed to see the regularity and the measure produced by a truly insignificant and lawless body. A paradoxical water clock where solidity displays its fluidity, the hourglass surely provides the first measurement of brief time. It is the glib symbol of a useless duration.

Powder, talcum, flour, ashes—all similarly hold the attention of alchemists and chemists in every period of the development of prescientific thought. It seems that a crushed body, in losing some of its individuality, simultaneously acquires an unexplained character of mystery. Powder arouses the suspicion of poison, it is an essence that, depending on the dose, may bring remedy or death. It is a sorcerer's material.

At times, it is due to the *uniformity* of dust that we think we can attribute a broad role to matter. Thus, a late-eighteenth-century author will associate dust with germinating soil. Air, says Deluc, works on terrestrial matter "ceaselessly and in a thousand ways. By simply rubbing all bodies of matter it removes such tiny particles that they are unrecognizable. The *dust* in our dwellings may well be an example. Whatever the nature of its source material, it is a grayish powder that seems to be everywhere the same. The formation of *germinating soil* is probably related to that. All surfaces of the earth, from the hardest rocks, the most arid sands and gravels, even metals, suffer the *gnawing* action of air; and their particles, reduced, decomposed, and reconstituted in myriad ways, are likely the main source of germination."[8] And so this uniformity, advanced on the basis of our inability to discern specific characteristics, is enough to explain that dust properly encompasses the most varied vegetative needs. In other words, vegetative comparison is no better able to discern differences between grains of dust than is human sensory activity. It would seem that, as solids diminish in scale, they are substantially simplified and thus become *elements* suited to the most diverse constructions. These particles, adds Deluc, "extracted or fixed by procedures that bring them closer to their initial elements and, in our eyes, cause them to take on the same appearance ... are thus suited to spread in the seeds of plants, to expand their tissues, to take on all the properties that characterize each species, and to maintain them as long as the plant exists. After the plants are destroyed, these same particles take on the general character of *germinating soil*, that is to say a ready-made reserve for *germination*."[9] Let us also note, in passing, the paradoxical idea that dust, the final result of all destruction, is easily posited as indestructible. The attribution of eternity to the atom in certain philosophical systems may have no other origin.

Thus, at the basis of our intuition of powder and dust are very curious judgments of value, since substances in this form are sometimes considered trash and, in others, worthy matter. We are amazed, in fact, when going from one judgment to the other. For example, who has not been struck to learn about new forensic tests? It takes all the talent of a Locard[10] to convince us that a criminal investigation can be explained through microscopic analysis. We had been led, through a pragmatism that was as crude as it was negative, to tacitly assume that substances lose their individuality when reduced to dust. We are therefore quite surprised to learn about the material individuality of the infinitely small. Moreover, thanks to the effortless dialectic of amazement, we are soon led to be amazed at our surprise. Thus, we don't hesitate to exaggerate newly recovered individuality and to postulate a set of qualities for material particles that are more characteristic than aspects of matter in its massive form. And so it is, as I will show, that naïve atomism assigns to elements qualities that are apparently not related to regular solids.

In addition, we might understand the influence of pejorative judgments often associated with dust by recalling certain related conditions such as *wood-rot* and *rust* that keep intuition in the prescientific stages. For example, *rot*, in and of itself, serves as an explanation, and the seventeenth century does not hesitate to believe in the action of a special worm that attacks metallic substances—dust from rust is considered the same as dust from wood rot. A *table of presence* might bring the two phenomena together and provide a Baconian explanation adequate for knowledge limited to relating two intuitions.

Along these lines, going on now to generalizations, we will understand that one of the great arguments of atomism, endlessly repeated by the various schools, has to do with wearing down the hardest of bodies. The temple's bronze doors hollow out under the faint touch of the hands of the faithful. The atom is now a worn solid. After a long success of creative effort, everything returns to the chaos of disassociated and mixed atoms. This theme of the general wear and tear of things, of the destruction of integrated forms, and of the amorphous mixture of diverse substances is the basis for numerous materialist philosophies that can thus adapt their pessimism to a sort of aesthetic decline of the Cosmos.

THE QUESTION CAN ALSO BE APPROACHED FROM ANOTHER ANGLE. IF DUST and powder are valued for their direct explanation, we will be led to value the pulverization of solid bodies as a truly fundamental process. We will not hesitate, at that point, to explain complicated physical phenomena in terms of the idea of pulverization, which will play the role of a *simple idea*. That is how Hélène Metzger quite rightly characterizes the psychology of a seventeenth-century chemist: "Like all dabblers in pharmacy (Arnaud, 1656) crushed solid bodies in a mortar. He believed that all chemical operations have some relationship to that one, that they may be finer or cruder, but that, ultimately, the chemist's entire art boils down to the mechanics of pulverization."[11] Pulverization is the clear and primitive idea, to which all chemical reactions must be brought back: "What is calcination? Seventeenth-century chemists reply that it is a process which consists in pulverizing different bodies by fire, either by action of the flame's actual fire, or by action of the potential fire found in acids and other corrosive materials."[12] In the *Encyclopedia* (under "pulverization")[13] one can also read that "calcination, either by fire or, by the assistance of niter and of sublimation into odors, is still, as to its effects, a type of pulverization." We can thus readily see that, for several centuries, the pulverization of substances was not merely a procedural means, but indeed had the importance, in the mind of the chemists, of a fundamental intellectual framework.

V

UP TO NOW, WE HAVE OBSERVED POWDER AND DUST IN THEIR RATHER DIMINished or at least static and inert aspect. But it is when we come to fine, light dust stirring and shimmering in a ray of sunlight that we really grasp the master intuition of naïve atomism. This is a spectacle we often contemplate in our reveries. It is capable of liberating our thoughts from the everyday laws that regulate active and utilitarian experience. In a way, it contradicts such willful experience, leading us to sever the link established by Bergsonian philosophy between our actions and concepts. Reflections born of this spectacle immediately have a speculative tone. They readily take on the function of learned reflection since they explain the general by the singular and the special, a method used more often than one might think at first glance.

The entire set of departures from usual laws, when manifested in the aerial play of dust, is precisely what makes its intuition so appropriate. The speck of dust, in particular, departs from the general law of gravity. For a truly primal intuition, need it be noted, it floats in a *void*; it follows its fancy. Of course, it responds to puffs, but with what freedom! It illustrates the *clinamen*.

Through a profusion of colors and iridescence, the speck of dust dancing in the light also illustrates the multiple properties of an isolated object. Upon looking at it carefully we think we understand that the element, simple in its substance, can be composite in its attributes and modes.

But the principal explanatory value derived from the speck of dust, its true metaphysical meaning, is surely that it brings about a synthesis of opposites. It is intangible and yet visible. A strange object that affects but one sense, that presents itself as a kind of natural abstraction, an objective abstraction!

But let us go further—in this experience what becomes visible is the invisible. In fact, as long as a reflected and diffuse light fills the room with a uniform clarity, the room is empty, the dust is invisible. Let a sharp, straight ray appear and immediately this ray of light reveals an unknown world. This is really the first experience of atomism; this is where atomistic metaphysics touches upon the basic physics of the atom; this is where speculative thought finds support in an immediate intuition. From now on, in fact, we can recognize our right to postulate matter beyond sensation since, in a way, experience has shown us the invisible. So we postulate the atom of matter beyond the experience of the senses. We are ready to speak of the atom of smell, of sound, and of light since we have just *seen*, in an auspicious and exceptional experience, the intangible atom of *touch*.

Such nimble and free matter might obey the impulses of the soul; it might be spirit itself. As Léon Robin recalls: "Aristotle, who does not name the Pythagoreans when he speaks of soul-harmony, expressly attributes only two opinions to them: according to the one, whose relationship with atomism he does not fail to point out, the soul is made up of dust particles floating in the air, highlighted by a ray of sun, and perpetually on the move, even in the calmest moments, while according to the other, the soul is seen as the root of their movement."[14] In both cases, therefore, there is a correspondence between the elements of the soul and the elements of matter. The atoms of the soul, adds Émile Bréhier in interpreting the same intuition,[15] are in equal number to those of the body and are juxtaposed to them by alternating one-on-one with them; they are constantly renewed by respiration. How can we not then consider that for early thought, the spirit of life takes shape in a puff of breath; how can we not relate the intuition of the mind to the observation of light animated by atoms that fill an infinite void?

From an animistic perspective a sort of passage to the limit[16] can be discerned that allows us to transcend matter. But in a more general and more material way, that is precisely where the epistemological usefulness of the observation of dust resides—it prepares and legitimizes a *passage to the limit*. That is the way Descartes makes use of this intuition in his book on meteors. In speaking of vapors and exhalations, he points out that specks of dust are much larger and heavier than the small portions that constitute vapors; nevertheless, he adds, "that does not keep them from pursuing their course toward the sky."[17] Here one can really perceive the powerful example of phenomena viewed in a ray of light. What dust can do, how could the atom or the fine matter of exhalation not be able to do? If dust manages to escape gravity, how could the atom not find its independence? If the experience of dust is still crude, all that is needed is to pass to the limit and we will attain, through thought, an atomic physics that will give the impression of being rational while still maintaining an experiential basis. Here then, in short, is the progression of arguments that carries forth the initial intuition and that establishes philosophical atomism as a doctrine at once rational and empirical.

VI

IN CONNECTION WITH THE INTUITION OF DUST, ONE SHOULD ALSO STUDY the intuition of the void, for it is not difficult to show that it also is a quite positive one. In fact, upon reading the Greek philosophers, we become convinced that the entire polemic over the void amounts to either aiding or combating that intuition. But in any case, when we first encounter this basic intuition, the void poses problems from a metaphysical perspective by the very fact that it raises no problem from a psychological point of view. Such a polemical outlook is well suited to demonstrate that the void and dust are truly immediate and important facts of experience.

This essentially derivative aspect of the metaphysical problem of the void is so clear that the problem is sometimes stated in a totally metaphorical, even unwritten way. We read in Aristotle, for example, that "if we are to believe the Pythagoreans, the void is originally found in numbers, for the void is what gives them their particular and abstract nature."[18]

All these arguments against the void are also interesting to my way of thinking in that they underscore the power of a first intuition carried into the most varied domains. Thus, for Plato and Aristotle, it is a question of combating the idea of a void that would be an instrument of general annihilation and that would bring to all substance the contagion of nothingness. They argue, in fact, that in the void, bodies would lose their specific properties. For example, with respect to motion, the void would erase individual dynamic properties. Thus, Aristotle concludes that "all bodies in the void would have the same velocity, and that is not admissible"[19] since the void would, in fact, take away from motion the fundamental Aristotelian characteristic of velocity.

Besides, in a more general way, the properties of bodies in Aristotelian physics are, as we know, entirely relative to their environment. A given property is more than *localized*, it is truly *local*. The attributes of a substance must be forced, in a way, to remain in the natural venue of that substance. Otherwise, the substance could not really retain its attributes, which would undergo a sort of metaphysical evaporation. In the final analysis, Aristotelian dialectic is led to replace the intuitive void by a space that, if not real, is at least necessary to assure that objects retain their real qualities. It is acceptable for the space to be empty of substance, but it must maintain a relationship to the substances it contains. It must *realize* the minimum necessary for the principle of sufficient reason to be applied. This point of view is very clearly summarized by Léon Robin: "With the venue deprived of all natural properties of location, what reason would there be, in fact, for a body to move in any particular direction? How to explain, as well, the accelerated motion that, to the contrary, a body displays when in the vicinity of its *natural venue?*"[20] But really, in implementing rational necessities, all we did was fill space with reasoning, and we still have to make the characteristics produced by the immediate intuition of the void reappear. Thus, Barthélemy-Saint-Hilaire underscores the dialectical character of properties attributed to space and bodies by Aristotle. Matter that fills space, he says, "is not such that it can oppose the least obstacle to motion, and motion occurs with such a constant and perfect regularity that, evidently, nothing troubles or hampers it."[21] But then, who does not see that positing metaphysical fullness amounts to attributing to it all the characteristics of the intuitive void? Fullness even has as its only function to maintain the properties of things, to bind attributes onto atoms in some way. The initial intuition has been enriched rather than impoverished; it remains whole. Once again, metaphysics has recovered what it had willingly lost. After a long detour, we must come to the conclusion that space is not a physical environment like any other, that it neither impedes nor produces motion, that it leaves undetermined all the reasons it contains for forecasting phenomena. Metaphysical plenty remains a physical void.

If the reader hesitates to follow me in this affirmation of the persistent character of the first intuition of the void, I have in reserve an argument that will otherwise answer an objection that is quite natural.

Surely, no one has failed to object that, in fact, the experience of the *void* for the ancients as well as for common knowledge is obviously erroneous since all the early physical experiments are carried out in air, with an almost total ignorance of phenomena peculiar to the gaseous state. We should then concede that the direct intuition of the void corresponds in reality to the experience of a physical state that, in itself, is well determined although poorly known. But an error of thought or expression has nothing to do with the truth of an intuition. What must be called the tangible perception of the void is closely linked to a quite positive observation.

Let us try to specify the experimental characteristics of this intuition. Air for the ancients was always the wind. In ordinary experience, if air is immobile, it somehow loses its existence. Wind is always a power of coordination. That is why the disorderly movements of dust in a ray of sun are not attributed to the wind. Here again, these movements represent an exceptional state, and, through a sort of dialectic, they display an ambient *void* as still another exceptional state. Immobile air is decidedly the intuitive void. It has no action, and it is the indicator of nothing, the evident cause of nothing. Accordingly, by taking the experience of aerial environment such as it appears initially in its general and simple aspect, it has to be recognized that this experience is well suited to providing a proper substitute for the void. In the final analysis it cannot be argued that the scientific error of an intuition destroys the power and clarity of that intuition.

THIS IMMEDIATE AND ENDURING CLARITY EXPLAINS THE DIFFICULTY brought on by the first scientific experiments following the invention of the air pump. By following these experiments over the course of the seventeenth and eighteenth centuries, we perceive the transition of an absolute and clear idea to a relative and confusing one. This transition was psychologically difficult and the idea of a *relative* void, so familiar to us now, was long a difficult idea to analyze.

Initially this relative void was taken to be essentially artificial. For a very long time, it was called Boyle's void, after the English physicist who multiplied the experiments. It was a technical state whose properties seemed as new as radium must have seemed at the beginning of the twentieth century. Considered to be a paradoxical state, it drew astonishing, extraordinary, and legendary observations. To give just one characteristic example, let us cite the claim of distinguishing between properties of the void when air is removed from a cubic vial or from a spherical one. With the action of the air pump, the first would shatter, the second would resist.[22]

Finally, more learned intuitions, based on the image of rarefaction developed through statistical analysis, very slowly began to help follow experiments in their particulars. These intuitions have profoundly permeated the culture of our time. We must forget them in order to appreciate the play of detail from the earliest intuitions.

To sum up, atomism is, first of all, a visually inspired doctrine. To ambient air corresponds a void of optic sensation. The material characteristics of gasses can only be understood through a scientific experiment, with technical means that are difficult to apply. Optical characteristics thus conserve a sort of natural explanatory value. Dust and void apprehended in the same glance truly illustrate the first lesson of atomism.

Chapter II

REALIST ATOMISM

I

TO THE SIMPLE AND CLEAR INTUITIONS THAT SUPPORT ATOMISTIC DOC-
trines from the beginning, one must add the help of a metaphysics that is
equally simple and direct in order to explain the pedagogical success of these
doctrines. That is realist metaphysics. Realism will indeed come across even
more clearly when it corresponds to a better-defined object. The atom, well
isolated in the void and assured of the immutability of its characteristics,
is easily taken as the archetype of an independent and immutable object.

Yet it would seem that this realist position, a particularly solid one over the
course of the development of systems, is not an original metaphysical doctrine,
at least as concerns classical philosophy. Historians of Greek philosophy agree,
in fact, that the atomism of Leucippus and Democritus derives from the Eleatic
school. Metaphysical meditation initially understood *being* in its transcendent
unity. Leucippus and Democritus are credited with preserving a certain unity of
being while fragmenting and dispersing it within space. Of course, Democritus
accepts the void, but for him, it does not reach the very essence of being. This
point is brought to light in a penetrating page from Léon Robin:

> In his discussions of their doctrines, Aristotle attests precisely and force-
> fully to what, actually, seems to unite the Abderite and the Eleatic schools.
> He [Aristotle] says, in essence, that the Eleatics, disdaining facts and *at the
> risk of flirting with madness*, proclaimed Being's absolute unity and immo-
> bility. Wary of their intoxicating logic, Leucippus makes concessions to
> sensory experience; he tries to preserve plurality and motion, generation
> and becoming. Yet he concedes to the Eleatics both that true Being is free
> of void and that without it there is no motion. Nevertheless, since the

27

reality of motion is accepted as fact, the void must constitute, in the face of Being, an equally real non-Being; and since plurality is accepted, that it must exist in the non-being of the void, and not in Being, from which it could not emerge. Hence, in his eyes, Being is an infinite multiplicity of masses that are invisible by virtue of their smallness. They move about in the void. When they touch they do not form a unit, but in joining together through such contacts they induce generation while their separation yields corruption. This means that Leucippus and Democritus both "coined" homogeneous Eleatic Being into an infinite number of clear-cut units, all *solid, indivisible bodies,* all *atom masses.*[1]

Likewise, Mabilleau wrote: "Leucippus presupposes Parmenides."[2]

To use Léon Robin's felicitous expression, we can say that, when Leucippus and Democritus coin Eleatic Being in this way, logical value is, first of all, entirely preserved, for early metaphysical atomism treats the logical subject from an individual perspective, just as Eleatism treats the logical subject from a universal perspective. Thus, we go from Being taken in its absolute unity to special objects whose individual unity is assured. In such a metaphysical deduction, the atom is therefore considered a logical unit before being taken as a material unit. That explains the purely logical *simplicity* of the Democritean atom.

But if the atom is so logically and radically simple, all atoms must be identical. Whatever might differentiate them must be sought in their connections, where a contingency would be at play that Eleatic thought finds incomprehensible. Léon Robin clearly demonstrates that identical atoms are a metaphysical necessity:

> Since atoms are sheer extension, replicated to an infinite number of instances, any property not contained in this fundamental essence of Being will be excluded from them by virtue of the Eleatic method. All atoms thus have the same nature, without any qualitative diversity, just as the Being of the Eleatics. They can no more be qualitatively changed than they can be divided, so that they are doubly imperturbable. Since non-Being is unable to give birth to Being, they are ungenerated, hence imperishable.[3]

II

UPON CONTEMPLATING THIS TRANSITION FROM ELEATISM TO ATOMISM, WE might think that metaphysical reflection is all that is needed to make one doctrine grow out of another. But let us consider a pedagogical question, a

secondary one, if you will, but one that will help us understand the real affiliation of metaphysical doctrines: What drove the metaphysician to consider
the idea of partitioning Eleatic Being? There is no doubt that the impulse came
from sensory intuitions. There is, in fact, no intrinsic reason to abandon Eleatic
thought once its pure simplicity has been understood. Experiential reasons
are the only ones that can drive us to the atom.

But how can we confine experience to a simple role of metaphysical circumstance? Inevitably, sensory intuitions were bound to spread their influence
by degrees throughout the entire philosophy of matter. A broad avenue of
induction extended the narrow path of logical deduction. A pervasive realism
thus followed the logical ontology of Leucippus. As I indicated in my introduction, Epicurus established, in a manner of speaking, *a naïve atomism
starting from a learned atomism.*

Besides, even if we were to admit that the thought process that leads to
partitioning Being starts logically, does experience not suggest an ultimate
limit for such partitioning? Mabilleau saw this very well: The indivisibility of
the Democritean atom "is physical and not mathematical, and . . . it is induced
from experience rather than deduced from a theorem."[4]

And so, in this unlabored thought, this intuition that reconciles geometry and physics, this philosophy so little concerned with separating logic
and experience, this doctrine more at ease with likelihood than with truth,
we have an atomism truly made to span the ages, to reemerge in the wake of
different philosophies and to adapt to scientific knowledge over the various
stages of its progression. Because of the Greek miracle, this naïve atomism
is not chronologically first. Yet it remains primary. In putting together my
table of possible atomistic modes of thought, I must therefore not hesitate
to upset the historical sequence and take as an archetype the most substantialist atomism.

Let us consider the atom most laden with substance, most rich in qualities. Indeed, it is when benefiting from the greatest number of qualities that
the atom naturally comes across as an easy means of synthesis. We can see that
it is easier to understand than many other elements. Psychologically speaking,
it is really the first. I will call it *the atom of internal realism* to distinguish it from
an element of atomistic philosophy that does not follow all the seductions of
naïve realism. I will call this latter philosophy *externally realistic atomism.* We
shall see that this latter atomism, unlike the former, entirely affirmative one,
relies on numerous negations. But this distinction will become clearer as we
go on.[5] Let me now attempt to characterize "coined" substantialist thought,
dispersed and multiplied into the various kinds of atoms.

III

LET ME SAY RIGHT FROM THE START THAT REALISM IS THE PHILOSOPHY THAT
evolves the least because it is the simplest of systems. It explains everything
with the help of a single epistemological function, namely, the direct reference of quality to substance. Once it is asserted that a body *possesses* a given
property, any subsequent question is seen as useless, or at least as derivative.
In such a philosophy, it would seem, we can make an immediate distinction
between what is real in a phenomenon and what is illusory! Thus, the various
problems of substantial or phenomenal composition that we will examine in
the next chapter would have the double characteristic of being derivative and,
from certain perspectives, filled with error the moment they are articulated.
Actual metaphysics and actual science would not be found in such problems
of composition; they would be found in the discovery of the only real link, of
the only decisive and primary epistemological function, which will always be
in the *bond* between an individual substance and its qualities or, inversely, in
the relationship established between phenomenal qualities and substantial
qualities. Mabilleau will end up treating this barely discussed ideal in the last
pages of his book on atomism: "Scientific progress consists of linking external
displays of matter to its internal constitution in order to establish, through the
interdependence of the two orders, the unity of the law without which there
can be no true explanation."[6] Thus, the interdependence of an order of profound, hidden, and substantial entities and an order of apparent and visible
qualities is posited as possible and even clear. On reading a statement like
Mabilleau's, it would seem that the epistemological function, which, in the
interest of brevity I call the *realist function,* corresponds to a self-evident idea.
Through a connection strong enough to prove all, it joins inner substance to
outward displays of the phenomenon. Realism would then not be limited to
asserting the reality of a phenomenon but would be reinforced by asserting
the reality of a substance. We shall see how easily atomism is founded on this
double realism.

But here we can posit a metaphysical objection, for which Hannequin
offers an excellent formulation:

> In fact, substance can be given the upper hand over its modalities only
> on one condition, and that is that they emerge from substance itself,
> that they unfold in line with its power to produce and create them. But
> who will limit its creative power? Who will direct it? Who will force this
> absolute, under what are considered rigorously similar circumstances, to

twice unfold identical modalities? And moreover, what can circumstances, eventualities, and conditions external to itself be for a substance that is self-sufficient and can be nothing but self-sufficient under penalty of not being a substance?[7]

Indeed, why would we not assign all diversity to substance? How could circumstances influence substance if substance does not contain the possibility of such influence within its own attributes? In the final analysis, if a substance is sufficient to explain a single quality, it must explain all qualities. Little by little, in following this path we arrive inevitably at explanation by singularity. And so, from assertion to assertion, realism, launched from a cosmos antagonistic to thought, becomes "thing-oriented" (*chosiste*), and from thing-oriented it becomes atomistic. For the atom is the thing-in-and-of-itself (*chose vraiment chose*), which resists analysis through a position that is completely and definitely objective. In other words, for realist thought, the real cannot remain dependent on the fragmentation we impose; fragmentation must therefore reach its limit. Only under this condition can the real be defined apart from our action on the phenomenon. If subdivision were infinite, substance would truly be a phenomenon since it would constantly be conveyed through the illusion of a composition. If, on the other hand, the unity of Reality were indivisible, Reality would come close to being nothing more than the idea of a reality. Substance and reality confirm one another in an intermediate position. Inevitably, a realist philosophy must become a realist atomism. *Atomism is precise materialism.*

FROM THIS POINT FORWARD IT SEEMS IMPORTANT TO UNDERSCORE, AT THE level of the atom itself, the development, or rather the deployment of realism. The atom, which, if it were to stick to its initial epistemological function, would only furnish reasons to explain the composition of phenomena, actually receives as inner qualities all the traits of a phenomenon. Examples come readily to the reader's mind:

Do we notice that things taken at the level of our experience display a cohesion that is by and large the phenomenon most obviously hostile to atomistic intuition? Right away, we attribute to the atom hooks that will allow it to fasten onto other atoms, while also transferring interatomic cohesion to an intra-atomic domain that finds itself, in turn, equipped with an essential cohesion. And so, on the one hand, the atom is given two qualities that are almost opposites: it is both isolated and attached. On the other hand, the phenomenal quality to be explained becomes a substantial quality and no longer needs an explanation since it is posed under the sign of realism. What is more, if we realize how naïve the image of hooked atoms is, we will simply say, along with modern science, that the atom has affinities. But this more muted realism is not more illustrative. At bottom, beneath the vague term of affinity remains the intuition of interatomic links. The very fact that the multiplicity of links does not provide a measure of a molecule's solidity proves how obscure and misleading the intuition of affinity is. Indeed, where the structures of organic bodies are illustrated as a chains, it is not always where a double link occurs that the molecule is the most solid.

But amid these problems of cohesion it seems that a deductive effort masks the retreat toward realism. Realism is surely at its most illusory when intuitions are closest to the senses. In fact, over the entire development of the doctrines, all facets of sensation are readily attributed to the atom. In their triumphant individuality atoms will be called upon, for example, to directly explain tastes, odors, and colors. Often, however, atomistic philosophy will have qualms. Then a sensory attribute will be taken as a sort of link between the atom and the sense organ. Sweetness, softness, bitterness will then be explained by inventing a geometric adjustment between the pores of the organ and the shape of the atom.[8] But sometimes the attribution of a sensory quality to the atom is more direct; it is swift, like a metaphor or a synonym. Hence a cold atom will have sharp points to account for a "nippy" cold, and a rough atom will be charged with explaining a harsh taste. Prescientific thinking is satisfied with the play of linguistic pairs.

But to really show how direct the fundamental intuitions of atomism can be, let me quote a page where the chemist Lémery[9] sets forth proofs of his notion of acid and alkaline particles. Duhem, from whom I am borrowing this citation, quite rightly observes that Lémery's theory and the philosophy of Epicurus and Lucretius are related.[10] But, once again, an immediate intuition, and not a philosophical tradition, is the basis for Lémery's conceptions.

> Since there is no better way to explain the nature of something as concealed as a salt than by attributing to its constituent parts features that correspond to all the effects it produces, I would state that the acidity of a liquid is found in pointed particles of salt in a state of agitation. I don't think it can be disputed that acid has points since all experiments demonstrate it and all you have to do is taste it to agree. For the stings it produces on the tongue are similar or quite close to those you would feel from some material cut into very fine points. But an indicative and convincing proof that acid is composed of pointed particles is not only that every acid

crystallizes into points, but that all dissolutions of different materials brought about by acidic solutions take this shape when they crystallize. These crystals are made up of points of different length and thickness, and this diversity is attributable to the points of varying sharpness among the several kinds of acids.

It is also due to this difference in the fineness of points that one acid will readily penetrate and dissolve a compound that another is unable to rarefy. Thus, vinegar can etch lead while nitric acid is unable to dissolve it; nitric acid dissolves mercury while vinegar cannot penetrate it, and so on.

As for alkalies, they are recognized by pouring acid on them, for, as soon as that is done, or shortly thereafter, a violent effervescence takes place that lasts until there is no longer a body for the acid to rarefy. Such an effect can reasonably lead to the conjecture that an alkali is a matter composed of stiff and brittle parts whose pores are designed in such a way that acid points, once they penetrate them, break up and push aside everything that stands in their way . . .

There are as many alkaline salts as there are materials that have different pores and that is the reason why an acid will make one material ferment but not another. For the right proportion is required between acid points and the pores of the alkali.[11]

I have quoted this page at length because I think it illustrates well the pure and simple passage from phenomenal to atomic property. It is particularly striking that the points of particles find their proof in the "points" formed by crystals; that pores are postulated in alkalies simply to receive the points of acids; that the *strength* of acids is presumed to correspond to the degree of "fineness" in the points. It can readily be seen that, for such an intuition, a particle of acid is a *thing* in miniature. We know this *thing* clearly only through taste sensation and our description of it is greatly reinforced by hypotheses that, as Hélène Metzger astutely remarks, are "at once precise in details and vague in generalities."[12] We call upon the mechanics of joiners and carpenters, and on the properties of levers, wedges, drills, and saws. Elsewhere, Metzger points out the minute scale to which images can descend when intuition proposes to understand the infinitely small. Here is how the chemist Homberg[13] presents the theory of oxidation of mercury:

We can think of particles of mercury, having become bristly when riddled by light, as if they were chestnuts encased in their bristly, green shells holding each other in place rather than rolling down an inclined plane, as they would if they were round and smooth. In this state mercury is no longer fluid, having changed into a red powder whose tiny grains, stuck together by their bristles, make up fairly hard, larger pieces and irregular shapes, just as bristly chestnut shells would if pressed against each other to form large, irregularly shaped balls that would hold together well. These bristly points of mercury, when exposed to fire over a period of time, grow in number and size, intertwining and upholding one another so strongly that mercury becomes as hard as a rock. And since the points that make each grain of mercury prickly constitute a matter that is discernible and has weight, mercury in this state grows in volume and weighs more than it did before being put to the fire, when it was still fluid.[14]

But one might recognize in this a very Cartesian tendency to explain everything in terms of extension and geometric forms. It would not be difficult to find pronouncements that are more substantialist and thus more gratuitous. That's what happens when the composite quality of the phenomenon is attributed to the simple nature of the atom. By reducing the scale, we have the impression that we escape the tautology forever ridiculed by Molière on the soporific value of opium.[15] We would then say with Voltaire, for example: "We admit atoms, indivisible and unalterable principles that constitute the immutability of elements and species, whereby fire is always fire, water is always water, earth is always earth, and the imperceptible seeds that form humankind cannot form a bird."[16] We thus return to Anaxagoras's homeomerous principle of "Everything-in-Everything," in which elements are at once specific and complex.[17]

Everything that enriches the atom with attributes thus depends naturally on realist thought. In other words, it is on the atom that best rests the epistemological function I have called the realist function, the one that explains a phenomenon by quite simply attributing it, as a quality, to the essence of being. The atom would then seem to be a solid and permanent source of this attribution.

But the problem, so well known in the familiar form I have just outlined, can be deepened and carried over to a more metaphysical terrain. Let me give a few examples of this transformation.

One of the most important metaphysical enhancements has consisted in taking the atom as a cause. This is a particular case of a general thesis whose development can be found in the criticism applied by Schopenhauer to Kantian philosophy. Here Schopenhauer proposes to demonstrate "that the concept of substance has, in reality, no other content than the concept of matter. As for accidents, they simply correspond to different kinds of activity and, consequently, the so-called idea of substance and accident is reduced to the idea of cause and effect."[18] In other words, from Kantian substance to Schopenhauerian substance we have the entire distance that separates kineticism from dynamism. Every substantialism must be paired with a causalism, and it was quite right for André Metz to bring out the causal character of Meyerson's realist philosophy. Now atomism is where a substance, concentrated on a narrow and precise area, is truly linked to its attributes; it is therefore at the level of the atom that a substance can most easily be taken as a cause of its attributes. The atom, then, is substance taken as a properly defined and properly assigned cause. In an action defined at the level of the substantial element, we can, as it were, recognize the causal atom. It is, after all, in this form that Cauchy[19] will write the fundamental equation of dynamics, starting with force applied to a point of matter.

Whatever may be the case, moreover, for such general metaphysical propositions, the passage from atom-as-substance to atom-as-cause can sometimes be found within atomism itself. This passage is often so furtive that it comes close to being specious. Thus, Lasswitz points out how, instead of resistance, we speak of the force of resistance.[20] Between resistance and the force of resistance there is unquestionably a gaping metaphysical abyss. Here an effect truly becomes a cause. Our whole inner intuition of effort is worked into the thing itself and, in a sense, substance is then experienced from the inside instead of being observed from the outside.

In the same way, we can shift from more or less *heavy* atoms, taken as a simple translation of our sensation, to a *weighty* atom posited as the cause of an attraction. We will encounter such an enrichment when going from Cartesian to Newtonian atoms.

In a similar yet clearly excessive fashion, certain philosophers would attribute freedom to the atom. In my introduction, I took note of this added feature of the atom in Epicurus. In coming back to it now, I wish to recall that the atom thus enriched can account at one and the same time for the absolute individuality of beings and the prodigious variety of phenomena. For Epicurus, variety is not a superficial game, it truly has an internal root. That is what Mabilleau rightly underscored: "The atom must therefore possess an internal and immanent power—one not subject to the sterile identity of mechanical laws and able to create variety by producing combinations, first in motion and then in being itself (by which I mean phenomenal being)."[21] Moreover, the powerful originality of Epicurus's atom is so indomitable, so impossible to erase through manipulations, that when atoms join together through chance or play, they constitute multiple worlds that are themselves imbued with contingency and variety. Émile Bréhier makes it very clear: according to Epicurus, there is "no reason why there should be only one type of world and that, for example, they should all have the same form and contain the same species of living beings. On the contrary, there are very different ones, thanks to the diversity of the seeds that made them."[22]

This exaltation drawn from freedom, and this individualizing force are all the more striking for initially having been concentrated in a smaller element. It seems as if the atom appropriates the mystery of freedom as it will appropriate the problem of life. There is, in fact, a kind of endosmosis between the concept of an atom and the concept of a germinating seed, and this fusion of two obscure ideas corresponds to a renewed enrichment of the atom of internal realism. The *genitalia corpora* of Lucretius, the *semina rerum*, are more or less clearly evoked throughout atomistic literature. Occasionally, the most ingenuous animist intuitions contribute a strange development to this thesis. Thus, in a 1674 study on elasticity, William Petty goes as far as to attribute sexual characteristics to atoms. According to Todhunter, he explains elasticity "with a complicated system of atoms to which he gives not only antipodal properties, but even sexual ones."[23] To justify his assertion, Petty claims that the passage from Genesis (1:27), "male and female he created them," must be taken to apply to the least of nature's elements, that is, to atoms as well as to human beings.[24]

Basically, the seed, obscure principle of the fate of being, represents, at least, the living being in its most condensed unity. It thus has become a model of atomic unity. With its deep structure, it simultaneously gathers all the qualities and all the development of the individual, even bringing about, as Alexandre Koyré has shown so well,[25] the metaphysical synthesis of the most flagrant contradictions: that of being and nonbeing, that of change and permanence. In the face of such contradictions how much more tolerable must seem the intuitive opposition between the atom's extension and its indivisibility! The seed thus provides one of the richest and best coordinated examples of the atom. We recognize that it cannot be divided without annihilating its functions. The seed gives proof, all at once, of its unity, its causality, and its life. It is the atom of life. If now we consider that animist intuitions can undoubtedly be concealed during certain periods, but that they are always ready to reappear, we can understand the misleading clarity that the obscure idea of a seed brings to bear on the geometric idea of an atom. The seed is an atom with an internal structure, one that can just as easily be interpreted as a root of diversity or of development. Thus, the entire phenomenon of being, in space as well as time, turns out to be explained by the seed.

Based on this metaphysics of the seed, we can now see clearly how realism posits the atom as a substance that really *produces* its attributes. Of course the most characteristic atomism will be the most immoderate. This is the one that, in all cases, always uses the same realist function, whether for attributes, forms, or accidents. Contrary to the ideal of modern scientific thought, it tends to reduce the *laws* of a phenomenon to the *properties* of substances. Making a system out of naïveté, this atomism even rejects the problem of phenomenal composition. It considers that there are no *composed properties*, or rather, that composition explains nothing. The entire value of an explanation consists in establishing a tautology that goes from substance to the qualities that characterize it.

IV

SCIENTIFIC ATOMIC REALISMS WERE A REACTION AGAINST SUCH IMMEDIATE attribution of tangible qualities. They worked to establish a scale of values for all of the properties and to determine the *fundamental* characteristics of the atom. Nevertheless, all atomisms that retain the realist function as a guiding idea, no matter how reduced that function's application, share a clear and undeniable metaphysical similarity. Before dealing with the problems of composition that are so important in classifying atomistic intuitions, it is fitting I think to point out several doctrines where the naïve atom becomes impoverished in attributes while conserving its richness as an essence, while still accentuating its substantialist value.

Since my intention is to be clear rather than complete, let me go right away to the most restrained atomisms, to those that might be called monotonous in the sense that they accord only one fundamental attribute to substance. Particularly instructive in this regard are the atomistic schools affected closely or remotely by Cartesianism. I have in mind here the theories of Cudworth and especially of Cordemoy.

For Cudworth, as well as for Gassendi in fact, atoms do not naturally constitute the totality of Being, as was the case for the doctrines of antiquity. For a Cartesian there is also "thinking substance." But this makes the qualitative poverty of atoms even more striking. As Pillon says: "the idea of spontaneous motion is not contained within extension, so it is not suitable for atoms. If they cannot move on their own, their motion must be transmitted to them, it must come from the outside."[26] Let us be clear, the term "from the outside" does not refer here to the action of another atom but indeed to an action that implicates a nature altogether different from the material one. The clash of atoms is then nothing more than the phenomenon of a deeper action, which through the intermediary of *plastic nature* goes back to God himself. The atom is then no more than a fragment of extension, no more than a geometric commentary on the impenetrability of bodies.

Cordemoy's thesis also accentuates this essential passivity of the atom and ends up with a veritable occasionalist doctrine of interatomic actions. Pillon summarizes this Malebranchean influence: "Not only is the principle of motion not within the atoms, but in no way do they act one upon the other and they cannot even be considered secondary causes of motion. Yet we believe and we willingly say that they transmit to one another the motion they received. But that is only an appearance. . . . It is God who causes motion to pass from one atom to the other on *the occasion* of their encounters."[27] Thus, the reciprocal action of clashing atoms corresponds to no profound reality. Such an action is a pure phenomenon; it cannot be *explained* by impenetrability since it is an error to attribute it to impenetrability alone. Divine action is necessary right down to the miniscule phenomenon of the encounter of two atoms, for God is the provision of all action. All else is form. We have indeed returned to a minimum and univalent atomism, although this univalent atomism is still thought of in terms of inherence. For Cartesian atomism, form is in fact really inherent to matter. Referring to Cordemoy's philosophy, Lasswitz recalls that it is because form belongs to substance that the substance of the atom cannot be divided.[28] In this doctrine the atom appears solidified by its geometric surface. This is not a contingent surface. It is not the mere limit of some inner effort at extension, the terminus of an internal push. It is really contemporaneous with the creation of being; better still, it is contemporaneous with the thought that creates being. That is why we can say that the surface of atoms is the geometrical site of their substantial qualities. This surface is really carved in intelligible extension.

In fact, just the idea of substance is enough to assure unity for the atom in Cordemoy's philosophy. This point is clearly brought to light in Prost's thesis: "Cordemoy states that if the atom is indivisible, it is because it is substance . . . he identifies, along with Aristotle, substance with unity. . . . It matters little that we differentiate . . . parts, their nature as substance maintains the unity."[29]

It is quite curious to note that, as soon as an atomism is outlined, a more realist reaction is in the making. Here we have, once again, the same dialectic as the one I pointed out between Democritus and Lucretius. It is interesting to see the same dilemma repeated between atomisms stemming from Cartesianism and from Newtonianism.

For an account of this enrichment, let us take the essay by Pillon that I used in reference to Cordemoy. Newton's fundamental contribution is the example of an action that puts the entire mass, and no longer just the surface, into play. As Newton's supporters put it:

> It is demonstrated that universal gravitation comes from a force that penetrates to the center of the sun and the planets, without losing any of its power, and acts, not according to the quantity of the surfaces of material particles, but according to the quantity of matter. Thus, no rotation, no known impulse, no mechanical cause, in the usual sense of the word, can be its source; hence, it is the effect of a force of attraction that is primordial and essential to matter.
>
> To this general attraction were joined others: there was one to determine the cohesion and hardness of bodies; then came magnetic attraction, electric attraction, chemical affinity.... What characterizes the philosophical spirit of science in the eighteenth century is the idea of the plurality and essential diversity of the forces, properties, and principles of nature.[30]

How better to characterize the return of realist thought to a philosophy of simple inherence? How can we not also see in such statements a resurrection of substantial forms and secret qualities? This is a criticism that was repeated time and again against Newtonian philosophy. While conceding that Newton, protected by mathematical safeguards, personally escaped the realist seduction, we must admit that the course of his doctrine was bound to place the root of all the external manifestations of the atom at the atom's center, as a quality of its inner substance. We return imperceptibly to an explanation by inherence.

This return to Aristotelian philosophical conceptions is especially apparent in a quality that most likely stems from sense intuitions, namely, hardness. Pillon's deduction is well worth considering in this case: "If the hardness of bodies, experiential hardness, comes from forces inherent to atoms, to what should we attribute the hardness of atoms? We will have to say that it, itself, comes from forces inherent to the parts that make up the atoms, and that the hardness of these parts comes from forces inherent to smaller parts, and so on. Or, if we want to avoid infinite regression, we will have to assign to atoms or to some sort of ultimate parts an essential and absolute hardness, similar to Aristotle's essential and absolute lightness, a hardness that is not of the same nature as the bodies, even though it is assumed as a result of the experience we have of the hardness of these bodies. We will then have to differentiate between two kinds of hardness in nature, where one, unexplained and mysterious, serves as a postulate in explaining the other. In a word, we suddenly will have to abandon the theories of modern physics to return to an idea from ancient physics."[31]

The relationship of atomism to Newtonian philosophy deserves much further study, of course. Metzger devoted penetrating pages to such a study: "atomism," she points out, "is implied in the law of attraction as understood by chemists." And she adds, in what is an important proof, in keeping with my thesis, of the imperative character of intuitions: "The chemists formulated this consequence of Newton's law as a first evidence, without discussing or deducing it."[32] We would find more highly instructive intuitive assemblages in other doctrines, for example, in the particle theory of light established by Newton,[33] but, as will be recognized, these assemblages always depend on the play of intuitions that we have identified.

TO CONCLUDE, I WILL LIMIT MYSELF TO INDICATING STILL ANOTHER PATH of development along which the atom was, in a sense, annihilated. In fact, we will see it lose, in favor of its dynamic value, one of the characteristics that, until now, was considered truly fundamental, namely, its very extension.

Explanation from within is what will spoil the conception of the atom as a figurative extension. Let me focus my discussion on an analysis of the principle of cohesion. This cohesion is due to a force of attraction among the component parts. The atom thus owes its existence to a mutual attraction of its parts. But then its parts have the same explanatory value with regard to the atom as does the atom with regard to the bodies that it constitutes. It must thus be admitted that intra-atomic cohesion postulates subatomic roots. And so, thanks to an internal property, we find ourselves led to endlessly segmenting the atom. In other words, by the very fact that it received all its properties from within, the atom cannot establish its individuality through a given figure that is completely external and carved out of extension. No link can be found to solidify figural extension and internal principles of cohesion. The method of explanation thus automatically shatters the atom as a means of explanation. As Pillon puts it so well: one must assume "either an initial extension that does not depend on cohesion and the force of attraction, or an initial force of attraction whose basis is not in an extended part, but in a mathematical point."[34] We then end up with an atomism where the internal and the external touch as it were; we have Boscovich's punctiform atomism.[35]

This time the atom is both intuitively and clearly an indivisible element; we have gotten rid of the internal contradiction to which we had been led by the need to give a variety of forms to indivisible atoms. But right away the difficulties expelled from the interior of the atom will reappear in the exterior. Indeed, the material point will, in a sense, have to defend its existence. We cannot imagine the contact of two points, let alone the clash of two points. For if two points were to touch, on purpose or by chance, they would coincide, and atoms posited as impenetrable would be mingled! Boscovich was thus led to postulate a repulsive force for minor distances while keeping Newtonian attraction for greater distances. We come to the realization that the description of phenomena henceforth entails the intervention of the geometry of space. It is from the relative position of atoms in space that all actions and, consequently, all properties of atoms derive. We thus arrive at a mathematical physics that strays from the traditional principles of atomism. We will not press our investigation any further. If we did pursue the relationship of doctrines along this line, we would encounter works of a mathematical order on point sets. A special study would be needed to separate and classify intuitive features of the problem of discontinuity as posed in set theory. From a strictly philosophical point of view, there is an interesting monograph that links Boscovich's intuition to modern philosophies of discontinuity, incorporating the theses of Cauchy, Herbart, Renouvier, and Evellin. That is the work of Nikola Poppovich.[36] This monograph is an excellent preparation for a consideration of the mathematical philosophy of Branislav Petronievics.[37]

Chapter III

PROBLEMS OF THE COMPOSITION
OF PHENOMENA

I

BY PLACING OURSELVES IN THE UNIQUE PERSPECTIVE OF THE REALIST PHI-
losopher, we have just seen that atomistic doctrines come in a considerable
variety. Depending on the number of phenomenal characteristics attributed to
the atom, all the cases of this kind can be classified into two extreme types: a
truly prodigal realist atomism that attributes all properties of the phenomenon
to the atom itself, and the most fully restrained realist atomism possible that
assigns one property as essential to the atom. Between these two metaphysical
intuitions, many intermediate forms are possible. This explains how a philosopher who subscribes to the realist view of the problem can maintain that
there is hardly "any theory more intricate than atomic theory."[1] This diversity,
which will seem even greater when we pursue our inquiry, entails modifications for certain philosophical problems. Thus, the question of phenomenal
composition is manifestly linked to the richness of the qualities attributed to
the essence of atoms: for prodigal realist atomism, *composition* has no meaning
since all attributes are postulated at the level of the atom, thus resolving the
issue of composition beforehand; on the other hand, this problem, so quickly
dismissed, becomes paramount for learned realist atomisms. Between these
extreme positions we can classify solutions to the problem of phenomenal
composition in order of increasing complexity, while an order of decreasing
intricacy would prevail for attributes postulated within the atom. Issues of
composition, combination, and synthesis are therefore narrowly linked to
atomistic intuitions. We need to examine them a little more closely.

Let us recall first of all a whole series of declarations that refuse to recognize the importance of *composition*. These declarations will seem less surprising to us now that we have seen that they are driven by an often tacit and

always unreasoned attachment to a naïve atomism. Thus Helmholtz, relying on the opinion of William Thomson, repeats that atomism can *explain* no other property than those attributed gratuitously and a priori to the atom itself.[2]

Berthelot offers a subtle characterization of this somewhat psychological need to deny the strangeness of a combination that produces brand-new qualities. Faced with a truly productive synthesis, "we would be inclined to believe in the intervention of some other component that analysis had been powerless to reveal."[3] In short, we end up making the power of combination substantial and seek a *compositional fluid*, an active element to explain the mutual action of elementary substances. It is an illusion, but it comes naturally to mind, and it takes numerous setbacks to weaken it. Berthelot continues by taking the example of sea salt: "still chlorine and sodium are clearly the only elements contained in sea salt. Synthesis has raised all manner of doubts in this regard, for it has established that chlorine and sodium can recombine, lose their qualities, and reconstitute sea salt with its original characteristics."[4] Thus, the idea that combination yields qualities is linked to long experimental practice. As long as we lack the means of verifying chemical doctrines through the two inverse experiments of analysis and synthesis, such doctrines remain on the metaphysical level; they develop in the realm of pure logic. And so they accept without discussion the principles of *total and perfect combination*. The following proposition is thus taken as a fundamental axiom, linked clearly to the logical doctrine of being: *what is in the whole is necessarily in the parts*.

Now and then this affirmation seems to play out on the very level of an ontological metaphysics. Paul Kirchberger, for instance, claims to escape from the propositions of Vaihinger and Otto Lehmann by taking fact and cause as realism. But, very curiously, he bases that realism on the following simple declaration: "In accepting Von Antropoff's principle: *if a body is composed of a certain number of parts, these various parts have the same degree of reality as the body in its entirety* . . . we find . . . an apparently stable and sufficiently firm ground to support the powerful structure of modern atomic theory."[5] Upon careful reflection, what we have here is a veritable ontological postulate—one that seems evident only because neither the perspective from which reality is examined nor what is meant by the *degree of reality* is specified. On the contrary, given a real characteristic, a degree of fragmentation can always be reached, leading to its actual obliteration. Crumbling to dust would be a familiar example of such a decline of concrete reality.

It is therefore difficult to reject the allure of immediate ontology; long experience with effective syntheses is needed to acquiesce to the *realism of synthesis* as opposed to the realism of the element. Thus Berthelot, writing on the atomistic doctrines of antiquity, quite correctly stated that these doctrines remain "alien to the idea of composition as such."[6] From a pedagogical perspective as well, it is always very hard to differentiate the basic intuition of mixing from the idea of combination. The best way to clarify this distinction is to define combination by the very fact that it creates radically new characteristics. I shall insist on this point.

II

THANKS TO SCIENTIFIC CULTURE, THE IDEA OF COMBINATION ENDS UP seeming simple and natural to us; but when we follow its development in science, we realize that it is surrounded by various intuitive nuances that render its conceptual precision precarious. Liebig understood the importance of such intuitive nuances very well. He takes a single, solitary fact and provides two articulations of it that may seem close but that, upon reflection, point to two different metaphysics: "Henry Cavendish and James Watt each discovered the composition of water: Cavendish established the fact; Watt had the idea. Cavendish says, water *is born* from inflammable air and dephlogisticated air; Watt says, water is *composed* of inflammable air and dephlogisticated air. The difference between these two declarations is great."[7] In fact, Watt is more advanced than Cavendish because he averts mystery. He accepts combination as a normal and clear fact. He understands that combination is enough by itself to explain the new characteristics of a compound. Cavendish implicitly and confusedly conserves the action of the life force, the intuition of development. For him combination is a birth, a creation that preserves its mystery.

We should also recognize that it was difficult to distinguish pure and simple chemical combination from various physical composites. This is a point I spent a long time examining with regard to Berthollet's intuition in a recent book.[8] I attempted to characterize the struggle between Berthollet and Proust by showing that the former incorporated a cluster of physical conditions into the chemical experiment that more or less masked the well-defined character of combinations. Proust, on the other hand, by insisting on presenting the phenomenon from its specifically chemical side, succeeded in showing that combination occurs without fluctuations, following rigorously general laws and that it is enough, therefore, to start from the idea of pure combination to define compounds completely. In other words, following Proust, the notion of composition reaches the rank of explanatory concept, and consequently this idea must be posited a priori as both clear and simple.

Yet we should not be fooled by this conception of perfect analysis, for Berthollet's intuition constantly gathers strength in our minds. Proust's chemistry is in fact a pure chemistry, a skillful and careful chemistry. Immediate facts, those that teach us realist philosophy, always include collaboration between chemical and physical phenomena. Gay-Lussac pointed this out in a simple and clear formulation: "Bodies in a solid, liquid, or gaseous state have properties that are independent of the force of cohesion; but they also have other properties that seem modified by this force, that is quite variable in its intensity, and, therefore, such bodies no longer follow any regular law."[9]

Since we are examining intuitions related to the composition of phenomena, we must therefore frame chemical combination within a margin that includes composites of physical origin. Within this margin must be gathered a whole set of concepts that go from simple juxtaposition to reciprocal penetration, from geometric to qualitative composition. But such concepts are often ill-defined, which is why chemical philosophy seems so imprecise when compared to mathematical or mechanical philosophy. To demonstrate that geometric intuitions are not absolutely necessary for an understanding of chemical combination, let me give the example of a highly regarded author who set out to link physical and chemical characteristics by using the simple idea of inner blending. For Sterry Hunt the typical chemical action is, in fact, to be found in *dissolution*: "Chemical union is an interpenetration, as Kant says, and not a juxtaposition as conceived by the atomistic chemists. When bodies unite, their masses, like their specific characters, are lost in those of the new species. Hegel's definition, that *the chemical process is an identification of the different and a differentiation of the identical* is, nonetheless, completely adequate."[10]

Moreover, we can follow, in several authors, the progressive loss in qualities when going from the chemical to the physical—as the composite becomes more physical, it becomes poorer. And since we learn mainly from large phenomena as well as physical ones, we always relate the power of a clear intuition to such poor composites. Thus Walter Spring writes in the preface to Hunt's book: "The combination of an element with itself, that is, the polymerization of a body, really has the effect of extinguishing its energy to render it unable to carry out certain functions. The chemistry of red phosphorous, simpler than that of white phosphorous, can be considered the chemistry of an inert body."[11]

In short, according to Hunt and Spring, a composite, when it is no more than a juxtaposition or an arrangement of elements in space, attains chemical indifference along with the perfect hardness of a geometric solid. Hunt concludes a synopsis on jade in the following terms: "The augmentation of density and of chemical indifference that is seen in this last species is doubtless to be ascribed to a more elevated equivalent, in other words, to a more condensed molecule."[12] This relationship between hardness and chemical indifference is also supported by research on silicates. What we see in this case, then, is a curious intuition where hardness is not primary but is acquired through progressive condensation. Thus, the idea of condensation, learned from everyday experience with mixtures and dissolutions, serves here to support intuitions that, at first glance, seem entirely opposed to it; proof that in the modern mind we are inclined to give a creative value to combination. We no longer feel the need to assign to the parts qualities that we observe in the whole.

III

ONE OF THE MOST HELPFUL PHENOMENA IN EXAMINING PHILOSOPHICAL issues of composition may be the compounding of the element with itself, as can be found in the case of allotropy.[13]

The fundamental philosophical question with allotropy, as Daniel Berthelot observes, is knowing whether the allotropy is physical or chemical.[14] This is not just a matter of words, and philosophers who, at this point, claim to base their arguments on the fundamental unity of science will inevitably fail to account for the profound and effective division of phenomenology.[15] In fact, physical and chemical properties are not approached with the same frame of mind. The problem posed by allotropy is thus philosophically complex. In order to explain how the same element, such as phosphorous, occurs with different *physical aspects*, is it really enough to link the *physical* idea of condensation, as did Hunt, to the concept of substance? Is it necessary to add a certain intensity to the substantial act in this way? Or, when dealing with allotropy, should we rather return to the dilemma that perplexes all of atomistic philosophy and choose between the following two modes of explanation:

Multiplication of atomic types—a multiplication so gratuitous as to lead us to postulate different atoms for the same substance;

Accentuation of the creative character of composition—an accentuation that would lead us to justify all of the chemical properties through combination alone, while denying chemical properties to the elements?

So, allotropy seems, in fact, to govern not only physical properties, such as solubility, color, and the system of crystallization but also specifically chemical functions. Daniel Berthelot writes that, according to Berzelius, "there are two parallel series of phosphorus sulfides, one in which this matter exists as white phosphorus, the other in which it exists as red phosphorus."[16] In some way color would thus be the sign of a deep differentiation whose origin can be traced back to the element. Marcelin Berthelot, as well, has shown the existence of special graphite oxides corresponding to the numerous kinds of graphites that are nevertheless all pure carbon.[17] And so we find a substance, taken to be pure and simple in its internal makeup, occurring chemically with a diversity that joins the purely physical diversity of the first aspect. There would thus seem to be a greater solidarity than ordinarily supposed between the physical and chemical characteristics of a substance. Hence, Marcelin Berthelot's study on sulfur comes to the following conclusion: there is a certain correlation between the function of sulfur in its chemical combinations and the forms taken when pure sulfur changes over to a solid state; where alkaline sulfides relate to crystalizable sulfur, oxygenated or chlorinated compounds correspond to insoluble sulfur.

What is more, a curious philosophical idea leads Berthelot to accentuate traits that point to the role of substance in a combination. According to him, "a great many facts concerning allotropy can be explained by a certain permanence of the properties of compounds, even in the elements derived from these very compounds. . . . It seems incontestable to me that several of sulfur's multiple states correlate to the kind of combinations from which they derive, or better yet, depend on two causes: the nature of the generative combinations and the conditions of decomposition."[18] And so, permanence of the simple within the composite—the ancient basis of all realism—is now opposed by permanence of the composite within the simple! The history of the combinations remains inscribed within the elements even as combination is destroyed! Having played a role establishes a quality—a far cry from quality unquestionably taking precedence over role. Role is normally a function of quality; now we have quality as a function of role! We discover from this problem a new reciprocity of substance to attribute that deforms traditional realism. It seems that in a way attribution may be substantially convertible, just like the logical conversion of predicates into subjects.

According to this intuition, chemical combinations will thus allow us to explore the degree of physical condensation. For Marcelin Berthelot, carbon's capacity to yield an extreme number of compounds relates to a more hidden capacity, one whose proof is already revealed in the variations of specific heat, namely, the fact that condensation occurs in a great many states! At its simplest state, carbonaceous material (which in all its natural varieties is virtually nonvolatile) would be, in a perfectly gaseous state, comparable to hydrogen. In fact, "a high temperature acting upon formene and benzine successively engenders carbides that are increasingly rich in carbon, decreasingly volatile, and have a constantly growing equivalency and molecular weight. We thus arrive at tarry or bituminous carbides, and finally at coals themselves. The latter still contain a few traces of hydrogen that heat is unable to extract. What is needed is the intervention of an element that craves hydrogen, like chlorine working at a red-hot temperature. Coals are thus not comparable to true elements but rather to extremely condensed carbides that are very poor in hydrogen. . . . Pure carbon is itself no more than a limit state that can barely be actualized under the influence of the highest temperature we produce. We know it as the terminal extremity of molecular condensations, and its current state is very far removed from its theoretical state, that of a gas . . ."[19] If we think about it, the entire perspective from which we habitually look for a purely simple element is what is overturned in such a conception. *A simple element would now always be a simplified element* that should be imagined not at the origin of a world but, indeed, at the end of a technique.

Whatever the status of Berthelot's theoretical perspective, the very fact that it can bring about an epistemological upheaval is enough to explain the muddled character of the problem of composition. Philosophers have not examined this problem by referring to the experimental program of chemistry; chemists, for their part, have not forced themselves to determine exactly where they make use of the philosophical function of realism. Many questions thus remain unresolved. Does the attribute derive from elementary substance? Does it derive from composite substance? Does it come into being at an intermediate stage? There is no doubt that in furthering these questions realism would split into several schools. Realism maintains its unity only at the cost of an imprecise application.

The problems of composition will become more precise only when a principle of *additivity* capable of accounting for experiments will be defined for each phenomenon. We will then realize that simple addition corresponding to elementary arithmetical intuition is no more than an abstraction, or at most a single case. When we examine a real object in all its aspects, we always end up finding cases where what is added is composed. At times the law of combination takes on a very general character. Thus, in relativity theories, thanks to the assimilation of the notion of mass and the notion of energy, we have come to consider, under the label of packing effect, a combination where the mass of the compound is smaller than the sum of the masses of the components.

Even at the level of elementary chemistry the principle of additivity would benefit from always being systematically defined. In his clear and original manual Marcel Boll does not fail to emphasize the importance of the problem. He singles out "*additive* relations, in which the successive addition of the same atom in a molecule produces constant variations in the properties of the body under examination . . . ," as well as "*constitutive* relations because they depend on the structures of the molecules. . . . Conversely, the term *colligative* relations is given to those that are merely the function of the number of molecules present, without taking into account either their structure or even their nature (the pressure of gases, the properties of solutions . . .), at least on first approximation."[20] Pure and simple additivity cannot, therefore, be considered an a priori imposed upon the experiment. One must always inquire experimentally if things are associated in an action like abstract numbers added to one another. In this regard, it is even a good idea to reject the evidence of immediate intuitions and to formulate a problem. We should not be surprised, for example, if a Réaumur should prudently inquire of an experiment "whether the strength of ropes surpasses the sum of the forces that compose these same ropes."[21] It seems that, by following intuitions of elementary arithmetic too slavishly, we neglect the exact mathematics of phenomenal composition. It may appear extraordinary, says Urbain, that we should fail to recognize a link between physical properties and the constitution of bodies beyond (*pure and simple*) additivity. . . . Is this from inadequate mathematical knowledge? It would be absurd, and even disrespectful, to assume that physical chemists know no other mathematical functions than those occurring in the form of polynomials."[22] And: "When the (*pure and simple*) additivity of a physical property is questionable, we accept that constitutive influences predominate. This is an evasion since these influences are themselves considered additive. The most certain thing that can be said in such a case is that systematization in the form of a polynomial is not applicable."[23] In other words, we can sense the emerging need for a mathematics of the composition of chemical phenomena.[24]

We thus come to a clear realization that modern science tends to free us from simple first intuitions. We must not be bound by any a priori view if we wish to confront all of experience. Problems of atomism will therefore gain from leaving behind the attraction of immediate realism. They must first be formulated as recapitulations of experience and then reconsidered within constructive thought, in which the scope and meaning of initial assumptions will be explicitly defined.

The analyses that follow will be given over to presenting this new aspect of atomistic philosophy.

PART TWO

Chapter IV

POSITIVIST ATOMISM

I

AROUND 1830, EVEN BEFORE AUGUSTE COMTE'S INFLUENCE COULD BE FELT, a truly positivist atmosphere begins to frame French scientific thought. Also starting at this same time, a reaction can be seen in Germany against adventurous speculations of which Hegel's philosophy had given too many examples. Paul Kirchberger, who makes this observation, sees here a reason why Berzelius's influence was limited, and researchers were diverted from investigating the electrical properties of the atom.[1] An electrically oriented study, by enriching the chemical atom with an additional characteristic, might have assured its realist value, but such a study was not carried out. Henceforth, and for a long period of time, the atomic hypothesis would be put forward as *solely chemical*, like a systematically unilateral synopsis of a phenomenon understood in only one of its particular characteristics. Besides, positivism is often satisfied with such a piecemeal phenomenology, one that sets forth an experimental design rather than a complete description of the phenomenon.

Thenceforth scientific atomism was taught under the protection of positivist precautions.[2] Such precautions, endlessly repeated, aimed to bring science back to a systematic phenomenalism by avoiding both realist declarations and purely theoretical ideas. In practice this systematically intermediate position proves very difficult to maintain, and so positivism is sometimes inclined toward realism and at others it relies on a rational organization of experience. It should not be surprising, therefore, that confusion is always possible on the edges of opposing philosophies. In particular it should be kept in mind that realist intuitions are always beneath the surface and that logical values remain readily seductive as well! It will then be understood that positivist atomism ultimately occurs, with polemical undertones, between the twin temptations of the real and the logical. Psychologically speaking, positivist atomism thus becomes so unnatural, so static, that it seems like a code of precautionary rules on avoiding error rather than a method of thought in search of discoveries.

The criterion of positivism is nevertheless clear: postulate nothing that cannot be verified in the laboratory. But the conditions under which hypotheses are experimentally verified do not involve the whole problem of the formation of hypotheses; positivism, reduced to its own doctrines, is quite incapable of coordinating a priori theoretical thought. From this point of view the use of the atomic hypothesis in the last century [the nineteenth century] is particularly interesting.

When one talked of the atomic hypothesis, one added that it was only a working hypothesis, a provisional assumption. At times one even professed the purest form of nominalism. The atom was but a word. Maurice Delacre still writes: "We use the word *atom* only as an empirical term, without any sort of philosophical commitment, and without making any conjecture on the constitution of matter."[3] Thus, the word *atom* should not be used at the beginning of an investigation, as a fundamental intuition, but we should come to it at the end of an experiment, like a useful summary that designates a particular aspect of the experiment. The atom would not lead to a definition of things but would be kept only as a definition of a word. Taken as a whole, atomic theory would, at most, be a scaffolding to bring experiments together, or even a simple pedagogical means of linking facts. Thus, Vaihinger cites Cooke's work (*La chimie moderne*, 1875) and adds the following very characteristic observation: "The book is founded entirely on atomistics, and yet the author denies that he is an atomist."[4]

In this regard it would be instructive to explore the mindset that governed the teaching of chemistry at the beginning of the twentieth century, and even just a decade ago in France. Most schoolbooks, following peculiar ministerial directives, postponed the atomic hypothesis until the end of the chapter on the laws of chemistry. Often the atomic hypothesis appeared in an appendix in order to emphasize that all of chemistry should be taught in good positivist fashion—through facts, and only through facts. One had to present the laws of proportional weights—laws that are simple, clear, and so well interconnected in the atomic intuition—by avoiding any reference to that intuition. It took skill not to utter the word *atom*. One always thought about it, but one never talked about it. Some authors, seized by late scruples, would give a short history of atomistic doctrines, but always after a uniquely positive presentation. And yet how clear these rigorous books would have seemed if we had been allowed to read them in reverse! How true it is that atomistic intuitions are available to us, that they come to the student's mind just as they came to the mind of ancient philosophers. All it takes for these intuitions to clarify and support scientific thought is to give them the right to be heard by assigning them a specific name.

Also very symptomatic was the debate tending to prove that Dalton had remained a stranger to initial intuitions.[5] With Dalton it was undoubtedly the first time that atomism was introduced into laboratory research; it was the first attempt to verify chemically a doctrine developed until then almost exclusively in the direction of geometry or mechanics. So, on this question, we must believe Dalton himself. An examination of his journal, says Kirchberger, leaves no doubt that he started with the fundamental intuitions of atomism and that these intuitions led him, as to a hypothesis, to the law of multiple proportions. Lange very ingeniously shed light on Dalton's intuitions by differentiating between the two otherwise quite similar methods of Richter and Dalton. Dalton's research "led him, like the German chemist Richter,[6] to the hypothesis that chemical combinations occur according to very simple numerical ratios. But, while Richter immediately jumped from the observation to the most general form of the idea—that is, concluded that all phenomena of nature are governed by measurement, number, and weight—Dalton strove to find a perceptible representation of the principles on which these simple numbers of weights and combinations could be based. And that is where atomism met him halfway."[7] By itself, Dalton's technique, a fairly crude one, could not have allowed him to move from phenomena to principles. Contrary to the positivist ideal, he went, instead, from principles to phenomena, from intuition to experimentation. But the positivist era was well under way. Very quickly the psychology of Dalton's discovery was reformed, and the history of science was written as if developed thought should respond to early thought and clarify its principles.

The teaching of chemistry and the history of scientific thought have now just gone through a period where artifice aspired to take on the value of a method. If one adds to this reason for confusion the fact that realist intuitions remain evident among the most careful positivists, one realizes that positivist atomism is fairly difficult to isolate. I will introduce it, therefore, in a way that is more categorical than historical in order to bring out as much as possible its distinctive characteristics.

II

IT IS OFTEN SAID THAT MODERN CHEMISTRY REALLY STARTED WITH THE
systematic use of the scale, by focusing on weight as the *only* criterion for the
scientific knowledge of substances. Here positivism can claim all the more
certainty for being more strictly applied. Delacre writes: "Weight is the first
principle, the most positive one, and if possible, it must be the only one."[8] It is
quite remarkable that we should aim at a quantitative identification of bodies
by focusing on a single quality that, in certain ways, can seem rather abstract.
It is particularly hard for intuition to accept the permanence of weight when
the volume changes; an often-difficult abstraction is needed to separate these
two characteristics. It is equally surprising that we should take the certainty
of a measurement as a warranty of its philosophical importance. Defining an
element through measurement, in fact, poses a philosophical problem that is
far from being resolved uniformly. Georges Urbain, one of the clearest positivists of our time, wrote powerful pages on this subject: "It is worth noting,"
he says, "that the method of the physicists diverges from that of the philosophers and mathematicians. The latter two start with an idea in order to
define magnitude. They consider that measurement should follow definition.
Physicists have a clear tendency to measure first and define later."[9] That would
amount to saying, paradoxically, that we don't quite know what we are measuring, but we measure it very well. In following the ideal of positivist science,
it would seem that science can be satisfied with a system of measurement, and
that scientific reality amounts to measurement itself more than the object
measured. Thus, Urbain continues: "It is interesting to note that if we define
magnitudes by the means through which they are measured, we avoid ideo-
logical pitfalls. Measuring atomic weights through chemical analysis and gas
density ensures an undeniable reality to the atomic weights. Even if the atoms
did not to exist, atomic weights defined by the way they are measured would
retain an eminently positive meaning. From the point of view that interests us
here, we can largely ignore the existence of atoms and declare with complete
certainty that atomic weight is an absolutely general elementary property."[10]
Schutzenberger had said the same thing: "By taking hydrogen, for example, as a
base and a unit, experiments teach us that the ratios of chlorine that come into
play in reactions are always whole multiples of 35.5 . . . 35.5 is what we call the
chlorine atom. We can see how narrow the meaning is. Beyond this definition
we are free to come up with whatever idea of the atom we wish; we can think
of it as an indivisible point of matter with its own real size and shape, or as a

particle, divisible itself in a certain way into smaller particles; we can admit that this atom has no real dimension so long as it retains a 35.5 to 1 ratio with the base; or still yet, think about it as the particular motion of a limited portion of continuous fluid that fills space. All of that matters little; nothing essential and truly scientific will disappear from the principles, laws, and deductions of the theory."[11] With respect to the problem that concerns us, the positivist point of view is thus expressed with admirable clarity. Philosophical intuitions are relegated to the rank of subordinate images. Atomic weights must be noted in the immediate phenomenon; they have absolutely no bearing on the philosophical problem of substance.

BUT WHAT MAY BEST DEMONSTRATE THE RUPTURE BETWEEN POSITIVIST atomism and the fundamental intuition of atomism is the very meaning that the wholly experimental concept of atomic weight takes on. In fact, if we confine ourselves to positivist epistemology, we are led to conclude:

1) that *atomic weight* is not a *weight*;
2) that *atomic weight* is in no way related to the *atom*.

The first assertion is evident when we note that atomic weights are expressed by abstract numbers and not by concrete ones as would be appropriate if we really were dealing with weights. As for the second assertion, it was hardly discussed in the nineteenth century, for all were convinced that we would never have any means of experimenting on the atom itself. That was also the reason atomic theory was considered the very model of the scientific hypothesis confined by principle beneath phenomenology.[12] In order to be quite sure not to give too realist a value to this hypothesis, the effort was made to identify the *table of atomic weights* as a *system of proportional numbers in combination*. And in this precise and clearly positivist designation every word should have its impact. At issue is a system, not just a table. At issue are numbers, not weights. At issue are entirely relative proportions, and not some sort of reference to the atom as an absolute of being. In other words, the concept of atomic weight is, from the positivist point of view, doubly misnamed. It is a concept that is based on an indirect intuition, yet it claims to render the experiment directly. An entire body of systematic experiments is needed to give experimental coherence to this concept. That really amounts to saying that this concept does not correspond to a determined thing as realist philosophy would have it. It may be no more than a symbol that lets us organize our experiment logically or economically.

Without feeling compelled to review the history of scientific doctrines in the nineteenth century—one can find this story repeated in many works—we must now endeavor to characterize what is both relative and coordinated in stoichiometry.

The most direct positivism—the most pure as well—could have been satisfied with the law of definite proportions advanced by Proust.[13] This law readily allowed us to catalog all chemical combinations by simply comparing two proportional numbers in combination. It seems, in fact, that we have in hand *the whole phenomenon* of combination when we know, for example, that the combination of iron and sulfur yielding iron sulfide occurs with a ratio of 56 to 32. With Proust's law we thus had the means to describe all chemical phenomenology, without subscribing to any theory or any intuition, by simply adding to every compound the proportional weight of its components. It should also be noted that it is to the compound and not to the element that doctrinaire positivism would link all the results of stoichiometry, since there is no a priori guaranty that an element will act in the same way in different combinations.

Yet success came from the opposite convention, amounting to assigning a specific number to an element. It became evident, in fact, that by taking into consideration three elements capable of providing, two at a time, three binary compounds, these three compounds could be described by attaching a single number to each element, and not two for each element as would seem necessary in describing the two compounds of which that element is a part. Thus, we immediately avoid even more hypotheses. Now the system of atomic weights is based on a kind of triangulation that seems to bear the trace of a profound ontology, and from which we could start a polemic between positivism and realism. It may be on this simple problem of proportional weight that the debates would be most useful. Let me, therefore, define the problem as clearly as possible, and for that, let us take an example.

We recognize, through analysis and synthesis, that a composite of hydrogen and chlorine occurs by weight at a ratio of 1 to 35.5. Similarly, a composite of sodium and chlorine occurs at a ratio of 23 to 35.5. These would seem to be two quite positive facts, without any theoretical or experimental connection. In particular, following the first two observations, if we come to examine the composite of hydrogen and sodium, it seems that knowledge of the two preceding ratios has in no way limited the essential unpredictability of empiricism. And yet we now observe that the proportions in a composite of hydrogen and sodium are exactly 1 and 23! We now have a sudden linking of facts, a cycle of being that closes with a perfection that we can rightly call rational, if we are willing to understand that the best sign of rationality is predictability. Starting with an initial wholly empirical link between A and B and a second wholly empirical link between A and C, we infer a link of B to C with the same assurance we would have if the link in question were an algebraic identity, yet without anything justifying this transitional method. To the facts is suddenly added a fundamental law. The axiom: two quantities equal to a third are equal among themselves thus becomes one of the designs of stoichiometry. This design is rendered substantively in Berzelius's law: if two elements join to yet a third one in given proportions, they join among themselves in the exact proportions in which they individually joined the third element.[14]

If now we consider that we will be able to take up the same problem with regard to a fourth then a fifth element, and so on, we realize that the set of proportional numbers in combination will be increasingly coordinated. Far from being lavish, empiricism will end up introducing a systematic frugality. Where we thought we were describing a collection of elements, we realize we are building a system out of substance. But perhaps we should be less amazed by the extension of the system than by the constant effectiveness of the same abstract number assigned to an element to measure its general bonding power. Even though we started from the most systematic positivist phenomenalism, we arrive, imperceptibly and despite ourselves, at realist statements. Indeed, how can we not say that a chemical element is characterized by an invariable number that is properly its own by virtue of its atomic weight?

We find a similar success of coordination on the subject of Dalton's law of multiple proportions, but this time the achievement is less surprising. In examining all the binary combinations of two elements, in the case where these two elements could yield not just one, but many compounds, Dalton recognized that the combinations, compared to the same weight from one of the elements, yielded, through the other element, proportional numbers that were simple multiples of one of them. A rather murky statement, but one that, in fact, would be clarified if we took the liberty to translate it into the language of atomistic intuitions! It has, in fact, often been said that Dalton's law could not have been perceived at the level of the experiment except for the small numbers that the chemical combination puts into play. If the compounds examined at the start had had the complexity of certain organic elements studied by modern chemistry, Dalton would have been unable to formulate his law; the margin of imprecision of the analysis would have completely muddled somewhat complex proportions. Proof once again that Dalton was guided by atomistic intuitions. It should also be noted that atomistic intuitions have always been developed in accordance with the clear intuition that arithmetic provides for small numbers. In certain respects, a fact that is expressed with large numbers always gives the impression of belonging to chance.[15] Nothing seems initially rational and clear but what is counted on one's fingers.

III

AND SO, VERY RAPIDLY, THE CHEMICAL PHENOMENON WAS PARTITIONED arithmetically and conformed readily to the atomic hypothesis. That hypothesis thus seemed to have completed its function. Yet, in fact, it was taken up again from many other perspectives. As we shall see, the attempt was made, following the same principles, to partition other phenomena beyond the chemical one. Furthermore, faced with the converging achievements of a single hypothesis, positivism finds itself confronting a philosophical question that it is hardly equipped to resolve: How can a simple working hypothesis succeed in such diverse areas? Is that not evidence of a value that is more real, or more rational? And consequently, because of its repeated and, in some ways, overly complete success, positivist inquiry finds itself facing a metaphysical obstacle: a simple hypothesis ought not be useful beyond its basic domain. To put it differently, if a convention succeeds from many perspectives, it is necessarily more than a convention.

Let us consider a few examples that will show how easily the atomic hypothesis expands; how the most diverse facts gather around it; in short, how its realism intrudes little by little despite all the positivist precautions.

One of the clearest cases is surely the law proposed as early as 1819 by Dulong and Petit relating to the specific heat of elements.[16] In their first treatise, Dulong and Petit began with a great effort at experimental precision. For eleven metals and two metalloids, they gave a table of specific heat carried to four decimal places, which may have been too ambitious an approximation. However, what must be well understood is that this table is drawn up outside of any reference to atomistic doctrine. If we examine the numbers in the table, we can find no connection among them. In particular, elements with very different densities, like gold and lead, have the same specific heat, within four decimal places; on the other hand, elements that have roughly the same density, like copper and iron, have quite different specific heat numbers. So we can see no connection between the density of elements and their coefficient of specific heat. It would appear, upon reviewing the numbers yielded by the laboratory, that we are faced here with an impenetrable empiricism.

But everything quickly becomes clear if we examine this table in light of atomistic intuitions. In fact, if we multiply the specific heat of each element by the corresponding atomic weight, we discover that the product is constant, whatever the element being considered. Dulong and Petit's law is found in this unexpected link.

When we look carefully at the twin play of approximations in the two tables, we realize that this wealth of decimals does not affect the constancy of the arithmetical product. We are then tempted to say that, even within the simple relationship of measurement, the two tables—one on specific heat, the other on atomic weight—though purely empirical at first, justify each other rationally. In any case, thanks to the use of the atomic hypothesis of chemistry in a problem of physics, it must be emphasized that arbitrariness comes to be limited. This feature had not escaped Dulong and Petit, who wrote at the very beginning of their paper: "Considerations based on the set of laws related to chemical compounds now allow us to formulate ideas on the constitution of elements that, although arbitrarily established on many points, nevertheless cannot be considered vague and completely sterile speculations."[17] Philosophers may not have pondered sufficiently this *automatic elimination of arbitrariness*, this natural and progressive establishment of physical rationalism.

Of course this relationship of reciprocal rationality between the two empirical tables offers a new means of forecasting and precision. Thus, when we find ourselves hesitating to assign certain proportional numbers to a compound, among which chemical analogies alone will not allow a choice, we will chose the number that satisfies Dulong and Petit's law.

Similar observations could be made about Raoult's laws that allow us to determine the molecular weight of certain substances according to the lowering of the freezing point of the liquid in which they are dissolved.[18] Here again, we would find coordination among various kinds of phenomena from the very fact that the atomic hypothesis serves to interpret their study. Thus, we have one more indication that this hypothesis brings clarity to an area where it is not an initial assumption. Strictly speaking, therefore, it no longer serves as a hypothesis in this alien domain. It is, in a certain way, sanctioned as a rational law through its success in a new empirical arrangement.

BUT THE MOST STRIKING EXAMPLE OF A HYPOTHESIS THAT FIRST BECOMES a positive and entirely phenomenal law, then finally a well-isolated fact, is surely found in the development of Avogadro's intuition.

Avogadro was struck by the very simple, almost rational character of the laws that Gay-Lussac had advanced for the combination by volume of elements in a gaseous state.[19] Thus, instead of the complicated, completely empirical weight ratio of 1 to 35.5 that determines the combination of hydrogen and chlorine, Gay-Lussac found that a volume of hydrogen joined to an identical volume of chlorine gave a double volume of hydrochloric acid. Gay-Lussac had also established that all gases have the same coefficient of dilation. In every way, the simplicity surrounding the development of phenomena related to gas led to the recognition "that there are also very simple links between the volumes of gaseous substances and the number of simple or compound molecules that form them. The hypothesis that first arises in this regard, and that even seems the only admissible one, is to assume that the number of integral molecules in any given gas is always the same, given equal volumes, or always proportional to the volumes." Avogadro explains his hypothesis in those terms at the beginning of his paper.[20] Intuitively, as Dumas emphasized, this amounts to assuming that "in all elastic fluids under the same conditions, molecules are situated at equal distances."[21] They are thus manifestly in the same number for equal volumes.

As for the effective determination of the number of molecules or their natural distance, that, of course, was a question that could hardly occur to Avogadro and Dumas. It was sufficient that the hypothesis should seem clear and natural and that it be used cautiously by affirming no more than the *proportionality* of the number of atoms to the volume of gas.

For a long time, there was an attempt to limit the teaching of chemical science to this simple declaration of proportionality. All references to the number of atoms were supposed to be eliminated, even despite Avogadro's initial intuition. One more example of positivism's effort to conceal atomism yet all the while utilizing its lessons! We then embarked upon the construction of very artificial concepts that seemed quite distant from any intuitive or pedagogical value, though they were directly connected to the facts of the laboratory. Outlandish expressions and verbal shorthand were invented that demanded long commentaries in order to be effectively understood. Consequently, even introductory books spoke in terms of gram-molecule, gram-atom, and gram-valence. And following this rational preparation, it was possible to state the very positive, very immediate *law* of Avogadro: for every gas taken at zero degrees and at a pressure of 760 mm. of mercury, a gram-molecule occupies the same volume, a volume of 22.4 liters.

In this form, the intuitive root of Avogadro's theory was successfully ripped out. No longer did we have a hypothesis but an empirical law, one that was accepted with its determinations by approximation and whose rationale required no further inquiry.

And yet, despite every prohibition, the life of first intuitions subsisted. Their use readily clarified the various derivative laws. Indeed, if we postulate:

1) that atoms exist,
2) that a particular atom has weight as one of its absolute characteristics, without the need to refer it to a second element in a possible combination,

then we infer from Avogadro's hypothesis that the density of a gas is proportional to the weight of the atom. From this flows an entire substructure that is clear and rational and that serves to support the assemblage of definitions whose conventional character was readily asserted by the former positivist method.

In fact, as we know, it is in favor of intuition and realism that science has progressed in this area. Jean-Baptiste Perrin[22] determined, in fourteen different ways, the number of molecules contained in twenty-two liters of a gas under normal conditions of temperature and pressure. Such varied experiments naturally yielded somewhat different results. But the convergence is sufficiently clear for us to be sure that we are not chasing a ghost. Avogadro's *number* is now one of the fundamental data of atomic science. We agree that in 22.4 liters of a given gas there are 60×10^{22} molecules.[23]

We thus succeed, along this path, not only in justifying Avogadro's hypothesis but also in taking something of the measure of it. Here a physics of the object is put back at the basis of a physics of the manipulation of phenomena. The line that runs from Avogadro's hypothesis to Avogadro's law, then from Avogadro's law to Avogadro's number, retraces the history of science of an entire century. Along this line an intuition becomes clearer and more precise. Such an intuition finally overruns positivism.

Generally speaking, the action of positivism comes across historically as an intermediary. It comes on the scene in contact with two different metaphysics, a rationalism of the hypothesis and a realism of convergent verifications. Two metaphysics? That may be the reason positivism claims to yield to neither one. In the realm of intuition, being free is to have two masters.

Chapter V

CRITICAL ATOMISM

I

HANNEQUIN'S INQUIRY INTO THE SCIENCE OF HIS DAY IS SO INFORMED AND so broad that it may seem unjust to characterize it rather narrowly as *critical atomism*.[1] Yet, even in its most cautious formulations, the great Kantian teaching reappears. Thus, at the beginning of his book—in an elementary precaution for any philosopher—Hannequin unhesitatingly poses the fundamental atomistic alternative: is atomism the result of the very constitution of our knowledge, or does it find its only reason for being in material nature? But right away, Hannequin realizes that the two alternatives do not carry the same weight; the second one, oriented toward realism, cannot be maintained absolutely and, consequently, cannot retain its characteristic function. Indeed, Hannequin, who here accepts as an axiom the fundamental principle of critical philosophy, immediately adds:

> In any case, even beyond its twofold appearance, it is still a single question in the end; thus making plain the truth that our mind cannot in some way disengage and step out of itself to grasp the reality and totality of nature.[2]

A critical theory of atomism will naturally look for a convergence of proofs starting at the very beginning, with the very first claim mind makes on matter. If this convergence within a priori thought turns out to be contrary to the heterogeneity of empirical data, such a thoroughly homogeneous convergence will seem all the more significant. We will now have proof that atomism does not derive from material nature but, on the contrary, that it comes from the world of apperception and intellection. From the very first page of his work Hannequin foresaw this characteristic *irregular application* of atomistic intuition. The atom of the chemist and of the physicist, he points out, have only the name in common. This observation—which, at first glance, would seem to please an advocate of *nominalist atomism*—leads us, if it is examined further, to pose the problem of *criticism*[3] in all its clarity.

First of all there will be no shortage of explicit declarations to justify the *critical* point of view: "Physical atomism is ... not imposed on science by reality, but by our method and the very nature of our knowledge; we would be wrong to think that it necessarily implies a real discontinuity of matter; it implies only that we make it so in order to understand matter, and that our mathematics introduces discontinuity while striving to construct it."[4] Once again, "atomism finds its justification in the very constitution of our knowledge." And, departing in this way from purely nominalist propositions, Hannequin adds that it will not be enough to show that, of all the hypotheses, atomism is the clearest, the most practical, and the most productive, "my goal is to show that it is a necessary hypothesis."[5] Now, such a necessity is one to be apprehended only in the realm of the a priori, in the very sphere in which *criticism* made its most durable discoveries.

II

INDEED, IT SEEMS ALMOST EVIDENT TO ME THAT ANY CRITICISM MUST BE directed by a structured arrangement and not be satisfied by some sort of *empiricism of reason* limited to finding and describing the laws that in fact guide understanding. A rule must underlie the law; on that basis alone can thought recover its unity even within the diversity of its own functions. Thus, when it comes to the atom postulated by a critical doctrine, it must be possible to identify a truly peremptory and primary characteristic. In fact, with Hannequin, such a decisive characteristic is not lacking. It is number. The atom, he says, has "its origin in the universal usage of *number* that marks everything it touches."[6] And later, "the atom ... is born of number; it is born of the need that pushes our mind to take its analysis deep into the regions where it will encounter the well-defined unit, the integrating, indivisible ingredient of which things are made, and then, not finding it there, determines, in this wholly ideal matter we call Space, the ingredient it is seeking and that it establishes there."[7]

In fact, if we really want to understand our author, it behooves us to discard right away the notion of an experimental and realist source for number. As Hannequin sees it, number is entirely a creation of our understanding. The unit itself remains relative to our action, perhaps to our will, or more precisely it is concurrent with our mental activity upon the world of representation. "Therefore, far from being drawn, through abstraction, from tangible and extended sizes, the unit is for us the only instrument that defines them and puts them at the mercy of our reason."[8] Of course, if the unit cannot be located in its entirety *within an object*, if the most it can do is to be predicated *in relation to an object*, this would apply all the more to the various numbers that always remain functions of the point of view taken by our understanding.

Objective plurality would thus be no more than a pretext to enumerate the acts of our understanding, the different stages of our knowledge. This plurality could always be increased in proportion to the refinement of our knowledge. The unit would thus be both a stopping point and a source of objective knowledge.

Must we take this claim at face value and see in the unit the *only instrument* able to bring about a preliminary determination in the world of the object, a determination that doubtless can be achieved in other ways but is nevertheless carried out on the first attempt within its own domain? Rational authority, if it could be considered equally decisive, would immediately bring about proof of atomistic criticism.

To clarify this question, it doubtless would be best to follow Hannequin's first efforts, at the point where he tries to install unit and number—which is to say, the atom of *criticism*—even within geometric continuity.

III

ONE OF THE HEADINGS IN HANNEQUIN'S BOOK IN WHICH HE EXAMINES atomism in geometry is indeed quite characteristic: "The Virtual Presence of Number in the Geometric Figure." The fact that this virtual presence is obviously in the world of the mind emphasizes the *critical* character of first intuition. Our author's entire effort thus consists in showing that we understand the relations of figurative extension only through the intermediary of measurement. About the essence of figures, he declares, "we understand only what can enter into relationships of ratio or equality, what can be counted and measured."[9]

Hannequin wrote at a time when the aim was to base algebra and analysis on whole numbers. If these efforts had been realized, a rational substructure would have been given to the irrational itself, in the sense that measurement would have always been reducible to finite or infinite sets of whole numbers. Thus, under geometric continuity, mathematical thought would have recovered a Pythagoreanism made up of numbers and sets.

Nowadays that arithmetic base seems too narrow. Even in analysis, operational extensions lead to such deformations of the notion of number that the simple traits of arithmetic can hardly be recognized in the generalized numbers. Moreover, Hannequin's measurement-centered intuition gives short shrift to all the ordinal, projective intuitions that, as early as the nineteenth century, had caught the attention of numerous geometers. On this point, then, Hannequin's essay comes across as quite artificial.

But this artificial character naturally presented no problem for a follower of critical philosophy. Quite the contrary, one recognized here the atomizing action of thought. The example was all the more striking in that regular and uniform extension seems, at first, not to provide a single pretext for an atomistic intuition. Let me then fully emphasize that atomism does not reside in the examined object, and therefore is in no way realistic, but that, on the contrary, atomism is linked to the method of inquiry. In fact, the unit comes about as a result of the equality of two extensions, with each of the two extensions, conceived as being equal, functioning with respect to the other as a unit. It is by setting up a connection—entirely dependent on thought—that we recognize a characteristic of comparative sizes. No prior realism can induce and sustain the intuition. It is not by considering size that we can understand a unit, but by giving it a function as a unit, by committing it completely to a synthesis, by taking it, if need be, as an instrumental unit, with an eye, of course, to examining a relation. All these statements indicate, I think, the *critical* meaning of numbered atomism, where the method of measurement overlays the extended formlessness rendered up by initial geometric intuition.

We find the same artificial character with respect to continuity understood as the development of quantity. Hannequin returns to the time when differential calculus was discovered. He characterizes this calculus as the means of measuring the gradual variation of figures. "By penetrating so far into the inner essence of figures, analysis was going to retrieve, or better yet, bring to bear in its wake, conceptual precision, which is the basis of all analysis, with its finite values and discrete quantity implying the notion of a constituent unit and indivisibility."[10] As can be seen, the word *retrieve* is immediately corrected by the expression *bring to bear*. The mind retrieves only what it actually contributes. Our understanding always *brings to bear* its precision on intuitively indeterminate continuity. Once again we have a synthesis, a construction that seems based on *indivisibles* in the sense Cavalieri used them.[11] We have a desperate attempt in which Hannequin accumulates the collective resources of an axiomatic and critical philosophy, hoping and doubting at the same time that continuity possesses indivisibles, but always limiting himself to proving that it receives them. It would thus be infinitesimal analysis conceived as a complex of relations that would "necessarily lead the mind to postulate indivisible components in every geometric object. Not that such components might exist: yet how else, other than by assigning an objective quality to geometric intuition and a reality to Space that neither can possibly have, can we assume the existence of absolute units that make up the figures? How can we expect, given a wholly ideal object, entirely made up of images and concepts, to understand the essential conditions of being, a true and absolute being? But precisely because it is ideal, when postulating the indivisible it is easier for our understanding to encounter it at the end of an analysis as a preexisting thing, than to introduce and constitute it by means of analysis itself. Whether with the infinite or the indefinite, our mind is geared to understanding only finite qualities. That is why, when focusing on an object, it calls such qualities into being so that the object may become intelligible."[12]

The problem would thus be to reconstitute with static elements the indefinite flow of change and to arrive in this way at a discontinuous theory of the derivative. This is a traditional debate that may well be eluded but not elucidated. Concerning this debate, Couturat directs a penetrating criticism at Hannequin's thesis.[13] In essence, on this very point, Couturat objects that mathematicians suddenly accept change within quantity. With the derivative they even set out to compare two rates of change, one of the differential of the variable and one of the differential of the function. Now, if we take each of these rates of change as independent facts, we notice that we have before us plain Change [Becoming], nondifferentiated, as meager in its principle as Being grasped by a concept. The comparison of two rates of change thus would not lead to a way of individualizing them.

Yet, by developing several obvious themes in Hannequin's philosophy we could answer Couturat's entirely mathematical objections. Of course Hannequin fails, like so many others, faced with the task of depleting continuity. But it does seem that he saw how much *comparison* was actively involved in the two processes of vanishing that contribute to the value of the mathematical derivative as a limit. It is at the very moment of comparison that change in the function is grasped as a function of change in the variable. It is through comparison that the quantities placed in relation to each other break up, atomize, and disperse. We thus encounter a source of *atomism through relation*, far from any ontological characteristic, and entirely justified by the hypothesis of *criticism*.

Moreover, we fully realize that this intuition requires us to utilize as many atoms as there are possible quantitative comparisons. The atom of quantity is therefore not a thing; it is the indicator of a role. This is the prolixity Hannequin has in mind when he praises Hobbes for having taken as an explanation of all forms of phenomena "as many kinds of fine matter, as many kinds of atoms as necessary, atoms that relate to each other the way infinitely small elements of various consecutive orders do."[14] It is certainly the case that if different atoms are needed to explain heat, elasticity, electricity, and chemistry, a truly innumerable plurality is needed to represent every point of our approximations. The more approximations are pursued, the smaller the atoms of the quantity in play. Such atoms are thus a function of method. They clearly fall under the sign of a critical philosophy.

IV

THE SEARCH FOR THE UNIT AND THE ELEMENT WITHIN GEOMETRIC CONTI-nuity seems itself to have been struck by an essential virtuality, and Hannequin finds himself obliged to conclude that "analysis that gives us the concept of a geometric element would never have given us the concept of the atom if our mind had not required a mathematical explication of nature."[15] All that remains of that entire search, in the end, is a confirmation that the *critical* position is correct. Indeed, the fact that the conception of the atom is *effectively prepared* by the conception of the geometric element shows that the atom is related to intuition and does not contradict the premises of the Transcendental Aesthetic. We can then turn to mechanics to complete "the definition that was only potential in geometry, so that atomism—whose hidden rationale goes back to analysis and, through that, reaches understanding—finds in mechanics alone, and in the mechanical explanation of things, the first cause and something like the motive of its appearance."[16]

In fact, this is now the decisive moment. This is the very place where—dealing with the principles of mechanics—the atom, as a factor of intelligibility, must demonstrate its success.

Indeed, we can say that with mechanics, "our mathematics gets as close as it can to phenomena and the real. Beyond motion, there is no more within them that it is capable of knowing."[17] Hannequin is writing at a time when *mechanism* is the very hope of science. For *mechanism* has a clarity that takes precedence over everything: a clarity of ideal and purpose. But now is not the time for us to judge that ambition; we only need to take note of its philosophical impact. Yet this part of our task is all the more difficult because in the very passage where Hannequin declares the *critical* direction of his own research, he recalls Kant's hostility toward any thesis of the discontinuity of mass. Despite that, the *critical* ideal is undeniable. For Hannequin, it is really a question of showing that in applying elementary concepts, developed on the basis of geometric intuition, we are compelled to postulate an element of mass, and that an identical epistemological *necessity*—an incontrovertible *critical* sign—leads us, imperceptibly, from the principles of geometry to the principles of mechanics. This is evidence, let me say in passing, that mechanics is taken here as a science of laws and not as a science of facts.

Let us then see how mechanics helps constitute atomism.

The first quantitative concept we come across when we want to construct a mechanics based on geometry is obviously the concept of *velocity*. Initially, it even seems that with this notion we can integrate the entire phenomenon of motion within the geometric intuition. In fact, on this precise point, all of Hannequin's comments are brief, which is explained by the fact that all of the commentary that ought to accompany a metrical comparison of time and space has already been developed in connection with the foundations of differential calculus.

Without following Hannequin, who would bring us nothing new on the philosophical problem of velocity, we might try to describe the relationship between velocity and the derivative. Basically, we could just as easily say that velocity is a derivative or the derivative is a velocity. From one statement to the other, however, there is an epistemological reversal, since the first statement brings the experiment back to geometric intuition and leads to understanding mechanics through analysis, while the second statement illustrates—if it does not explain—intuition through experiment. Now critical philosophy, by following the first thesis, will find it more satisfactory to understand velocity solely as a derivative. This is how we can more easily perceive time as independent from space, that is, in its mathematical role as an essentially independent variable. In this way as well, time is identified with inner intuition. Nevertheless, we must still be able to apply externally that form of inner sensibility that is time. For all its resistance to measurement, we must therefore find its measure, no matter how indirect it may be. Such a measure

time will get from space, from its relation to space, through the fully mathematical intermediary of the notion of the derivative that perfectly analyzes
the notion of velocity.

Perhaps this is the justification that could be applied to Hannequin's
thesis. Here again, atomism is constituted thanks to a relationship established between two essentially different processes of parceling. In itself, an
atomism of time is as inconceivable as an atomism of extension. But, *in their
relationship*, these two virtualities, so vague and obscure when separate, clarify
each other and, in the full meaning of the term, affirm each other. The twin
undetermined flow of time toward the instant, and of space toward the point,
designates through a simple relation a well determined limit. In other words,
the individuality of velocity—its realism—is shown by relating two incongruous atomisms, both marked by a fundamental virtuality. This relationship
is brought about, it goes without saying, within a critical thesis, through understanding that links in this way the two forms of sensibility.

Of course Hannequin could not have foreseen that a time would come
when we would be speaking of a real discontinuity for velocities and for
energies. His entire effort, when it comes to kinetics and geometry, consists in preserving the possibility of atomistic intuition; he attempts to show
that kinetic properties, for all their continuity, are not hostile to an atomistic inquiry.

This direction has thus taken us to the point where the atom must
suddenly be enhanced and truly constituted as a real unit. It is in the transition from phoronomy to mechanics as such that we must apprehend this
enhancement.[18]

In reading Hannequin, we realize that, beginning at this precise point
in his thesis, the atom, which up to now had been nothing more than a form,
is *henceforth taken as a cause*. So long as Hannequin is considering kinetics,
he remains in the realm of geometric thought; kinetics is then a sort of blank
mechanics, reduced to *effects*, reduced to phenomena. Thus, Hannequin
writes that the trajectory of a body in motion is "the geometric trace left in
space by the position of a body in motion; it is never the condition that makes
it one thing or another, be it rectilinear or curvilinear, or that causes it to be
described with a velocity that is at times variable and at times uniform. . . .
Phoronomy . . . is nothing more than a geometric language where the conditions and the laws of motion that are the focus of true mechanics or dynamics
are expressed by their effects."[19] This clearly acknowledges that a purely
descriptive science would not need to consider the point of motion in its

concrete aspect or active role, or, to put it differently, that the atom does not really crop up in the phenomenon of motion. But as soon as we seek to elucidate *causes*, the atom now takes shape and, in a manner of speaking, solidifies. A point with weight is postulated as the cause of phoronomic effects. A whole new metaphysics opens up, in which it will often be difficult to maintain in all their clarity the principles of *criticism*. No longer, in fact, can we remain within a pure and homogeneous relation, as was still done when exploring the geometric and kinetic conditions of motion. And since the domain of homogeneous relation is transcended, we see the realistic mode expand and consolidate. Once again, we have the usual temptation to posit the real beneath the convergence of relations.

Furthermore, the about-face is unequivocal: "Mechanics requires that we pay attention to the object in motion, whereas phoronomy considered only the trajectory. The body in motion is in fact the first condition of motion: in short, it is what moves, with its successive and changing positions tracing the trajectory as if it carried within itself the power of motion."[20] So many phrases that return us too swiftly to the absolute of being and thus stray from *critical* postulates! With respect to this particular conception, I must immediately put forth my own observations if I wish to preserve for the intuition of critical atomism its clear and simple meaning and its true metaphysical function. Accordingly, I develop a whole series of objections in what follows.

V

TO BEGIN WITH, THERE IS NO CONDITION THAT CAN BE CONSIDERED *FIRST* because there is no condition that can be considered *unique*. Even if a singular condition had any meaning, it would not inform us; it would not be productive of thought. By providing everything at once, it would provide nothing, for it would run counter to the very function of thought that must always acquire or rectify. It would not be possible to combine this initial condition with a secondary one because essence and detail must not be confused. Metaphysically speaking, it is always useless to back up an effect by the power to produce that *sole* effect. In short, knowledge must always begin with multiple conditions.

More to the point, in the classical definition of mass we can immediately recognize a reference to duality that excludes any primary status for one characteristic. In fact, in order to speak of a *mechanical cause* both mass and field must be present; one is not more *real* than the other. We cannot detach

force from mass so as to see stripped-down mass. From the moment we carry
out an experiment with mass, it is active, a force reveals its action. If the atom
is a cause, that is because it is not alone, because it is engaged in a complex
of conditions.

Hannequin's quite dense statement must immediately be disputed here.
He states that in passing from kinematics to dynamics "the body in motion
becomes a matter of inherence,"[21] whereas, in my view, we should limit ourselves to saying that the body in motion becomes a *matter of coherence*. Clearly,
Hannequin's philosophy hastily seeks out the object. "Unless mechanics has
no object, or what amounts to the same thing, that such an *object entails no
fixed determination that is subject to laws*, modifications of motion, and, consequently, modifications of the state of a body in motion could not occur without
conditions that must be sought out and that are the very object of dynamics."[22]
Should we not be concerned about the floating meaning of the term *object*
used three times here? If we give it its full meaning of *concrete object*, as seems
warranted by the passage I have highlighted, we see that Hannequin has just
yielded, suddenly and surreptitiously, to the seduction of realism. He accepts a
simple idea from which realism derives all its power: the law must necessarily
be the sign of a reality, just as the attribute is the sign of a substance. Taken
here in its elementary form, the full gratuitousness of that reasoning can be
seen: we come to believe that the point of motion includes, as one of its properties, the cause of its trajectory. And this belief is so secure that we do not
hesitate to reverse the argument and to pass from a trajectory examined from
the point of view of kinetics to the affirmation of a *real point* that produces the
trajectory rather than just traveling along it.

The cause of acceleration is affirmed by Hannequin in a way that is just
as specious. When velocity varies, he says, "the same reasoning requires us
to think that variations of velocity point to the conditions that affect change,
irregular variations for irregular conditions, and constant variations for persistent and fixed conditions."[23] This is a chain of reasoning that counts on an
absolute space. However, the mere relation of a body in motion to its trajectory
can indicate no more than a *relatively* accelerated motion, and the question
remains open as to whether the cause of the acceleration comes back to the
point itself or to the system of reference. But just the fact that the method of
reference should intervene in effectively determining acceleration clearly
shows that we will never be able to define the mass of a point by relying only
on the acceleration with which it traces its trajectory.

In fact, the step-by-step construction of the real as undertaken by
Hannequin is plainly linked to the order he follows. Thus, it cannot claim to

recover being as an absolute. Nor can Hannequin develop, as I think would be appropriate, a correlated theory of being, precisely because he posited certain geometric conditions as fundamental. "When all of mechanics sinks its roots into geometry, what abstraction could then vehemently separate the extended body in motion from the resistant body in motion, the *massa extensa* whose essence is to occupy a place in space from the *moles dynamica* that presupposes position, under penalty of being incapable of motion? ... A body in motion must thus be a whole in space for it to be a whole that serves as a divisor of force; it has to be a *volume* for it to be a mass, and the geometric measurement of the one has to be the standard of measurement for the other. Thus arise in our space portions of extensions that are initially simple figures, but once dynamics clads them with inertia, they emerge from the void that surrounds them as masses and become bodies. In a word, matter is thus defined as a volume whose dimensions are always specific, whose inertia expresses its whole nature, at least from the perspective of our mechanics."[24] This, I believe, draws all of critical philosophy in the direction of geometric knowledge. It sees the necessitarian aspect only in mathematical development. In my view, this method avoids a more properly metaphysical solution. But no doubt I should pose the question in clearly metaphysical terms.

As I see it, the problem should be formulated as follows: How can the same subject have two predicates; how can a single substance appear in two independent attributes? Let us see what light Hannequin's ideas might shed on this?

Must we simply equate the geometric and dynamic attributes and say, along with Hannequin, that "the quantity of matter or of mass is thus nothing more than the inertia of a full volume?"[25] Is that not to forget that the words *full* and *empty* have no meaning in geometry? Whether we like it or not, such equalization of geometric atomism and mechanical atomism cannot take place without support; we always come to an underlying individuality that by its very imprecision is ready to accept various forms. Thus, the phenomenological equality of mass to volume cannot be complete and pure; even when it is affirmed within a logical framework, we see the first signs of deep being, a trace of a poorly exorcised realism, that serves as a tacit hyphen. Hannequin will indeed write that "the quantity of mass is, in mechanics, always proportional to the volume it occupies"[26]—and with this we can hope to have all the laws of the logical transfer that runs from the geometrical to the mechanical—but the author finds it necessary to invoke immediately "the physical reasons that will give to elements in actual nature, multiple and varied densities."[27] In other words, my great hope would be that volume and mass are proportional, that to reflect on one is to reflect on the other, and that geometric *criticism* is immediately transposed into mechanical *criticism*. But we are immediately faced with a rebellious fact: the factor of proportionality between volume and mass is *density*, left over from an unreflected a priori empiricism. We will have to wait for the theories of Relativity for this density to be subjected to an internal reasoning; until then it will only be grasped through a thoroughly external reasoning, in its simple role as a factor of proportionality.[28]

But then would our intuition not gain in clarity and fecundity by joining the realm of pure logic? Would it not be best to remain within an integral critical philosophy? And for that to happen, would it not be necessary, with respect to the problem that concerns us, to institute a deliberately external coordination between geometrical and mechanical atomism? Since all bodies act in the same way in the presence of mechanical forces, how can we not see that mechanical forces are not absolute entities, but rather modalities that are the mark of a correlation from the outside, a correlation that is irrevocably involved in our system of geometric references? Hannequin looked at this problem with a remarkably penetrating eye. How well we sense, through the abstract enchantment of his style, what is mysterious in the conformity of logic and fact! "Volume, being a finite figure, is unconcerned with our measurements, and remains, without contradiction, an ideal whole, without parts, a merely possible number, made up of units that have nothing absolute about them and are arbitrary. But let it be clad with the resistance and motion of mechanics, and, through the irrevocable relationship of two concepts, let its extension be no more than the design and geometric support of inertia, and the elementary unit, which shortly before was merely postulated and perceived as potential by infinitesimal analysis, will impose itself as the condition for the existence of a finite mass."[29] We only have to be told why the relationship of two concepts is indissoluble and a new metaphysics is born, the logical impulse extends beyond its bounds, and a whole science is established that treats *logical equivalents* in roughly the same way that we determine *equivalencies* among the various forms of energy. Thinking is renewed because it can be transposed. In any case, we have, on this very point, the fundamental problem of the real: How do two conceptual domains, the geometrical and the mechanical, come, through simple superimposition, to suddenly take on the consistency of the real? How do two atomisms, both proceeding by arbitrary fragmentation, and in a totally ideal way, end up in some way resisting each other and thereby succeed in suspending the arbitrary and opposing the real to the idea? That, it seems to me, is the most logical form of the metaphysical question posed by Hannequin's philosophy.

VI

IT WOULD BE MUCH LESS INTERESTING TO FOLLOW HANNEQUIN IN HIS inquiry into the role of the atom in the natural sciences. On the one hand, the successes of atomism in chemistry seem assured; on the other, at the time Hannequin is writing, these successes are seen as carefully limited to their positivist qualities. Therefore, Hannequin's metaphysics is somewhat hampered, and his entire effort consists, first of all, in steering physical and chemical problems back to mechanical forms. Along this path, the *critical* tendency is not forgotten, however. It is even possible that the *critical* character is reinforced by the very fact that, through parallel paths, by similar methods, we find atoms that differ according to the area in which the atomistic doctrine is applied. Hannequin develops an entire paragraph (pages 145 and following) to show that "by way of multiple regressions the individual natural sciences lead to atoms of different and descending orders." He adds, page 147: "As numerous as the forms of the problem may be, the method does not change: it always comes across as an effort of our mind to replace the rich variety of living nature with the homogeneity of a matter that is lifeless and almost without attributes, where everything stems from motion and returns to motion." How better to state, first of all, that atomism is formulated not as a question related to the object, but rather as a question related to method, and secondly, that the central and really the only focus of the argument is located, as I have indicated, in the very transition from geometry to mechanics?

VII

THE POINT OF VIEW OF CRITICAL ATOMISM IS ALSO ADVANCED WITH GREAT clarity by Lasswitz. During his long investigation into the history of atomistic doctrines, Lasswitz was struck by the constantly dogmatic tone of these doctrines. And so, one task seems to him to remain. It consists of separating atomism from its usual dogmatic ground. This he undertakes in a small supplementary work: *Atomistik und Kriticismus* (Atomism and Criticism) (1878). He announces, as early as the preface, the results of his philosophical research in this area: "Given the fact that it conditions the form of our experience as a subjective factor, the nature of our sensibility requires us ... to choose a kinetic atomism as a theoretical basis of physics."[30]

This declaration places us immediately at the center of the polemic. The fundamental thesis is as follows: the kinetic atom is necessary for the *scientific* use of our sensibility. The atom is thus seen as less immediate than a given form of sensibility, yet as more than a simple rational hypothesis. The atom corresponds to particular assumptions that science needs to make in order to account for certain empirical results, assumptions that are more than necessary and sufficient hypotheses because they are needed to carry out intellectual functions. But then, as Lasswitz observes, the contradictions "that are readily found in the atom disappear in the face of critical thought, in the same way as the contradictions that had been noted for thousands of years in the essence of space and motion."[31] Atomism might, in a way, be renovated by the Copernican revolution of *criticism*. Although constructed by attempts at scientific understanding, the atom might still benefit from what is immediate in the a priori forms of sensibility. In reading Lasswitz, it would seem that the concept of the atom might well be upheld, as if the same necessity were found in geometric construction and in the material elements of that construction, as if the synthesis of space and substance were rendered by an a priori synthetic judgment.

On this path, we are inclined toward an epistemologically dynamic *criticism* that might admit of an evolution and a teleology in reasoning. In fact, Lasswitz does begin with this Kantian statement: "The possibility of experience in general is at the same time the universal law of nature, and the principles of the former are the very laws of the latter. For we know nature only as a set of phenomena, that is, of representations within us, and we can, therefore, extract their connection from nowhere but the principles of their connection within us, that is, from the conditions of linkage necessary within a consciousness, a linkage that constitutes the possibility of experience."[32] But Lasswitz makes this inaugural and fundamental possibility in Kant a fluid one. It then becomes a question of finding the conditions that make possible a particular experience, rather than experience in general. These conditions are no doubt still a priori because they are sine qua non conditions, but they are in some way dependent on their result. This *criticism* corresponds to a reciprocal correlation between theories and facts, a construction that is quite close to the axiomatic construction that I will be discussing in the next chapter.

Whatever this fresh nuance contributed by Lasswitz's atomism may bring to critical doctrine, this interpretation will serve to shatter the following traditional objection: atomism can only be a provisional and, consequently, arbitrary position on the problem of substance because, in assuming an atom whose dimensions are not determined, we allow the possibility of resorting, when the need arises, to a smaller atom. Here, then, is Lasswitz's answer: "But that happens only to the extent that a future experience is apt to discover facts that we do not yet know. It goes without saying that an empirical science is only required to explain what is known to it, and we have already acknowledged several times that determining the size of the atom is considered simply a task of the science of empirical knowledge (*and not a task of science itself, as realist doctrine would have it*).[33] Future centuries may find it necessary to take a step further than we have—but the question of the size of atoms remains a practical question; it is not related to the principle of knowledge. Only the characteristic of atomic explanation affects that principle, and that characteristic remains the same as long as the human organization remains the same. We specifically assert the following: the science of a particular period must end—or more precisely begin—with a defined group of atomic systems that can be thought of as encased one in the other, and that science must explain, starting from there, all there is to explain."[34]

In short, the concept of the atom is necessary from the *critical* point of view, in the sense that there would be no room for objective knowledge without the location of an absolute center to support contingent relationships. However, the set of essential and absolute properties is very quickly analyzed. We must then go on to empirical determinations of the concept of the atom and, right away, we realize that we are leaving behind the usual *critical* conditions. It is just such a segmentation of phenomenology that Lasswitz offers in the following passage where, depending on our sensibility and our own way of understanding, "there must be a phenomenal object that, in itself, is immutable, impenetrable, and very small, and that, as such, becomes the source of all change in nature. But then all the properties that we must necessarily attribute to the atom (*if we study it*) are depleted without regard to its connections with other atoms. . . . All other properties of the atom are properties of *atoms* [plural], in other words, all the other properties are conditioned by the connection of atoms."[35]

This split in phenomenology, for the benefit of properties that are in some way absolute, is extremely risky. Of course, in atomistic doctrines we are used to making an essential distinction between the properties inherent to the atom on the one hand and the properties that result from the composition of atoms on the other. But how should this distinction that was expressed with naïve intensity in realism be legitimized in a critical philosophy? We can ask, for example, how smallness can be attributed as an absolute to the atom, how the macroscopic experience of impenetrability can take precedence over an experience as simple as *mixture* and thus become essential to the atom, and how an a priori synthesis of the characteristics of impenetrability and immutability can be brought about. All are objections that could be reproduced, with the support of critical philosophy itself, against all attempts to determine a priori characteristics of an experience that is always apprehended a posteriori.

FURTHERMORE, LASSWITZ DEVELOPS HIS THESIS IN A VERY PARTICULAR WAY. He starts with the following proposition: "The critical theory of matter is necessarily a kinetic atomism."[36] But he does not maintain the primacy of the visual for intuition; in fact, he determines that the dynamic aspect of experience is entirely dependent on touch. "If it were only through the sense of sight that we knew the form and perception of space, we would no doubt possess a phoronomy, but we would not have any sort of mechanics."[37] And Lasswitz immediately adds, as if it were a self-evident consequence: "there would, of course, also be only an idealism, but no *criticism*."

These remarks, taken in themselves, are of a singular profundity; they seem to me to shed a great light on the classification of philosophical doctrines. In particular, we have here one of the reasons that could, it my view, clarify certain relationships between *criticism* and *idealism*.[38] What is active in critical philosophy is opposed, in fact, to a more passive idealism, where the conditions of knowledge arise more or less without a struggle, without having to omit anything from a crucial alternative because knowledge and mind are in total communication. In other words, for critical philosophy, experience is really a mental action, and without such an action, experience would remain an undetermined form. Even when taken at the level of sensibility, *critical* investigation must be an active investigation that goes beyond visual contemplation. This is all the more reason why it is impossible to *judge* the world of representation without intervening, for our concepts are diagrams of intervention and summaries of verifications.

In certain respects, it is the experience of touch, by forcing us to reflect on our visual experience, that determines the systematically reflective idealism that is *criticism*. The resistance that things put up against our necessarily unitary action is what leads us indirectly to attribute a unity of action to isolated objects. The atom is then naturally postulated as an active unit. It is less a unit of an indestructible figure than an essential unit of force, and it is toward Boscovich's intuition, already encountered at the end of chapter 3, that the metaphysical investigations of Hannequin and Lasswitz return us.

Boscovich's intuition could thus serve in some ways as a hyphen between realist atomisms and critical atomism. It is quite striking, furthermore, that this intuition of a punctiform atom, the root of key forces, should be directly utilizable by mathematical physics. Boscovich's philosophy does seem to rely on a minimum of assumptions. That may be what makes it suitable for relating to metaphysically diverse doctrines.

Chapter VI

AXIOMATIC ATOMISM

I

WITH DIFFERENT ORDERS OF MAGNITUDE, IT IS APPROPRIATE TO APPLY DIF-
ferent philosophical principles and a different language; for if measurement
may seem an essentially relative procedure, it is not a foregone conclusion
that all measured magnitudes are affected *simultaneously* by the same relativity. In other words, a complex of magnitudes is a positive characteristic of
a particular object, and nothing allows us to assimilate objects taken from
different orders of magnitude. To put it immediately in philosophical terms,
I would ask in what way and under what conditions the infinitely small is considered an *object*.

Indeed, the most striking epistemological characteristic of atomic science
is perhaps that it amazes us. We are not familiar, nor do we become familiar,
with the infinitely small. Often the only way to comprehend it is by distorting
our ways of understanding, in an entirely reflective activity, through a wholly
polemical use of reason.

Our language itself has taken its roots and its syntax from the world of
things and of actions related to our common experience. Our dictionaries
and our grammars are, at bottom, no more than *lessons of things*. Faced with
the infinitely small, it would seem to be enough to redefine our terms. But the
trouble goes deeper than that, for now the whole perspective of the definition
has changed: *where ordinary science is based on objects and seeks principles,
atomic science posits principles and seeks objects*. In the latter case the definition
of entities must thus remain entirely preliminary; it must then demonstrate
its productiveness in a domain that, ultimately, is no longer its own, in the
domain of common experience. Thus, we can see that, when studying the
infinitely small, we can no longer give a *definition that describes*; at most we

85

can only give a *definition in order to describe.* In other words, we must accept the definition in order to understand the theory and the facts, and not simply understand the definition in order to accept it.[1] The postulated atom is thus intimately opaque; only its role can become clear. And so, through linguistic necessity, the science of the atom now takes an axiomatic form.

Has it always been thus? Certainly not. What was missing for the atomisms of past centuries to deserve being called axiomatic was truly real movement in the epistemological composition. In fact, it is not enough to postulate, with the word *atom*, an indivisible element in order to claim to have put a real postulate at the basis of physical science. We must use this hypothesis effectively the way geometry uses a postulate. We must not limit ourselves to a deduction, often an entirely verbal one, that draws consequences from a single assumption. On the contrary, we should find the means to combine multiple characteristics and, through this combination, build new phenomena. But how could such a production be possible when, at most, we are focused only on proving the *existence* of the postulated atom, on reifying an assumption? The philosophical theory of the atom closes off questions; it does not suggest any.

No doubt it will be objected that a single nuance separates the atomism I call axiomatic from positivist atomism. In fact, in one or the other we find the same prudence; one, like the other, develops under the cloak of the traditional formulation: everything takes place *as if* the atom existed. However, such a comparison remains silent on a criterion for classification that I consider fundamental in epistemology, namely, the very direction taken by our reasoning. Indeed, the positivist school uses atomistic models as summaries rather than as principles. These models are more "as ifs" of expression and not "as ifs" of discovery. We should not be surprised then if the positivist models remain disjointed, if they are cast aside, or if they are restored when needed for pedagogical explanations. In the final analysis it is simply a matter of applying useful metaphors to describe immediate experience more or less clearly.

The principle of taking a hypothesis as a postulate also diverges from the classic doctrine of scientific hypotheses. To be sure, that doctrine has taken quite varied forms, whose details can be seen in a particularly rich chapter of André Lalande's book on experimentation.[2] But contemporary science adds a new feature to the "tradition of hypotheses." In fact, an axiomatic method must prove its value not only through its experimental results but also by the very movement of the thought that drives it. And this role is enduring, in the sense that the deductive ideal is never fully achieved; the structure remains incomplete by virtue of the fact that the basic postulates retain their independence. In other words, modern atomic science refuses to eliminate hypotheses completely; it does not simply attempt to join two descriptions of common experience; it aims to preserve the rational link that was used to pass from one experiment to the other. It seeks to *reflect on* the experiment by keeping in mind the postulates of the experiment. For example, what would be the value of an entirely phenomenological description of the ionization of gases without the theory and permanent image of the electron? Should we limit ourselves to seeing in the ionization of a gas a method of discharging a condenser? That discharge, on the contrary, is a simple sign indicating a hidden phenomenon whose process is in effect studied by the mind of the scientist. When following the movement of the lighted indicator on the graduated scale, the observer is thinking solely within the framework of the atom. Modern atomic science is dependent on its technical thinking and not on our common experience. That is why fully epistemological conditions must now be embodied in the preparation of an experiment. A particular experiment is henceforth entirely tied to a theorem. As such, it must be given a precise place in a grouping; it is a consequence and it has consequences. When epistemology further captures the attention of philosophers, we will realize more fully that *the order of ideas energizes ideas* and that it is through the order and composition of ideas more than through the analysis of ideas that thought can lead to discoveries. The architectonics of the science of the atom thus goes beyond the positivist realm. An undeniable solidarity joins thought and experiment in contemporary science, to the point that it is impossible to tell if the design of the atom is a map or a plan, if it derives from a descriptive science or a technique.[3]

In any case, it is appropriate always to keep in mind the many hypotheses of modern atomic science and to become acquainted with its particular methods of calculation and research if we want to understand fully its systematic quality. Such a pedagogy separates us from the arbitrary conventions accepted by positivist science. Without worrying about repeating ourselves, we can thus examine thoroughly the axiomatic character of contemporary atomic science.

II

ALL THINGS CONSIDERED, CRITICAL THESES PREPARE US BETTER TO ACCEPT the axiomatic direction of atomic principles. It even seems that, in its haste for construction, the mind is ready to accept as an *element in itself* any representation that is integrated as a whole and all at once into a structure. As Lasswitz indicates: "When thinking evolves to a certain degree, we do not feel the need to base certain simple representations on a deeper foundation."[4] But Lasswitz sees in this tendency the simple effect of an indifference born of habit, whereas, to follow the axiomatic ideal, we should clearly resolve to choose the basic element. Axiomatic thinking teaches us, in fact, to bring analysis to an end because, at most, analysis can do no more than prepare a synthesis. The epistemological function of the atom is to construct the phenomenon theoretically. We are justified, when engaged in thought, to treat as an element that which functions as an *element in a synthesis.*

Of course, in following this path, the element is integrated into a synthesis only as a result of its well-defined functions. Nothing vague must henceforth be taken into consideration in a postulated atom. Such an atom is the symbol of a definition, not the symbol of a thing. And this principle of holding exactly to what is defined, without ever going beyond it, is at the very core of the axiomatic method. In following this ideal, Lasswitz summarizes, at one and the same time, the character of an atom whose meaning is entirely specified and that is postulated solely for a synthesis: "Naturally, atoms will provisionally have nothing more nor less than the properties that are precisely required for the construction of a particular body."[5]

Moreover, to really constitute an axiomatic system, it is not enough to refine one by one all the basic definitions and to explicate fully all that is contained in notions taken individually. It would also be necessary to draw up a complete table of guiding principles. Here, mathematics could serve as a model for the physical sciences. Writing at a time when there was little question that mathematics and mechanics were deductive sciences, Lasswitz nevertheless does not hesitate to draw them closer to inductive physics: "Physical science and mathematics both have certain principles whose basis is rooted in our very nature and, for this reason, are immutable. But in mathematics, with its principles ... a complete table of definitions is also provided; such a table is missing in the physical sciences. Definitions in the physical sciences can only be obtained empirically. But the goal of science is to restore fully this table of definitions, and, if this goal could ever be attained, we would then have the possibility of treating deductively the whole of physical science as has already become possible for certain parts of this science."[6]

No doubt this ambition of total ascendancy over the real may seem poorly suited to furnishing a specific program for a science. However, we realize that such an ambition is far from illusory when we follow the efforts of a science that makes use of precise instruments. It is a distinctive trait of modern atomistics to generate a set of specific instruments, so that in some ways, it might properly be given the overall name of *instrumental atomistics*. With these instruments it is not a question of mentally reconstructing the muddled and vague phenomenon offered to our senses. Quite the contrary, the goal is a precise, schematic phenomenon immersed in theory. Not *found* but *produced*. Modern science tends increasingly to become a science of *effects*. These *effects* are given the name of their inventor. We speak of the Zeeman, Stark, Compton, or Raman effects. By contrast, never is a new chemical element given the name of the chemist who first isolated it.[7] Contemporary physics is thus in search of operations determined by theoretical perspectives. It is not led by an analytical intuition. It does not follow a descending but an ascending path. More often than not, we look for the *effect* without it having been previously provided by the experiment. We must first construct this effect in thought, in order to actually produce it. Thus, we are often struggling with two deficits at the same time: a deficit of mathematical prediction and a deficit of instrumental precision. No longer can we sustain the old philosophical thesis of a phenomenon seen as unknowable, unclassifiable, and offered as a whole to our mind. Already, Claude Bernard was able to speak of an *active experiment* in which the scientist is an inventor and in some ways, as he said, nature's overseer.[8] But never have the technical conditions of scientific work been so methodically coordinated as in contemporary atomic science. Basically, no matter how delicate the operations may be that produce the *effect*, that effect must appear mathematically the moment all the required precautions are followed. Yet these precautions are *countable*. Their number is not indefinite. That is an important and a fundamentally special characteristic.

Most notably, this *body of experimental precautions* forming the basis of a technique of physical effect has an entirely different meaning from the *body of conventions* with which a pragmatic philosophy would seek to form the basis of a science of the phenomenon.

Thus, a fully passive empiricism, whose essence implies innumerable conditions, tends to make way for an active experiment produced with precision and without any possible aberration, as long as care has been taken to carry out all the directions in order, one by one.

This detour brings us back to a wholly subjective certainty. Claude Bernard, in a singularly penetrating passage,[9] pointed to the distinction between an always indefinite objectivity and a totally inventoried subjectivity. If the experiment becomes *our own*, we might hope that it will benefit from the confidence owed to the immediate character of an action taken with a specific goal in mind. This, in fact, is the way that the axiomatic in geometry multiplied the evidence, in a manner of speaking, by adding together objective clarity and the light of awareness. In all areas, the axiomatic is a growing awareness of the exact conditions of thought.

Accordingly, instrumental science is on the same level as the body of definitions. *An instrument, in modern science, is truly a reified theorem.* By taking the schematic construction of the experiment stage by stage, or again, instrument by instrument, we come to the realization that hypotheses must be coordinated from the point of view of the instrument itself; devices like Millikan's, or those of Stern and Gerlach, are conceived *directly* in terms of the electron or the atom.[10] The basic scientific assumptions we now make about atomic characteristics are thus not mere scaffolding. They constitute the very framework of our instrumental science. That is why Vaihinger's doctrine, otherwise so suggestive, does not seem to me to have identified the true role of contemporary atomistic designs. For Vaihinger the atom is not, strictly speaking, a hypothesis; rather, it would seem to correspond to a fiction.[11] Accordingly, as fictions, all the characteristics attributed directly to the atom should be eliminated as soon as they have accomplished their entirely intermediate function, exactly in the same way that the symbol of an imaginary quantity in algebra must disappear the moment the results are set forth. It is precisely because the atomist intuition will end up being eliminated that we can load it with contradictory characteristics. And that would be true even with respect to [other] intuitions. Vaihinger goes so far as to say that an intuition, even if it is materially false, often serves provisionally in place of an exact intuition. In my opinion, such a deliberately *factitious* characteristic translates poorly the *technical* characteristic whose importance I have emphasized earlier. Factitiousness may well produce a metaphor. It cannot, like the technical, furnish a syntax capable of linking arguments and intuitions. Moreover, as Vaihinger himself recognizes (768), even though we can speak of the play of imagination when dealing with atomistic hypotheses, we must at least recognize that this game is not illusory. Far from leading understanding toward error, this game facilitates its task.

THESE IDEAS MAY SEEM TOO GENERAL, AND IT MIGHT BE OBJECTED THAT ANY technique calls forth the same observations. Nevertheless, it did not seem to me to be in vain to show that *atomistics in particular has become a technique,* with its own instruments, methods, and experiments. Now any technique flows from numerous choices. It accepts, in certain ways, the ideal of the initial contingency of axiomatics. Above all, it must allow great freedom to preliminary intuitions.

But the axiomatic aspect of atomic hypotheses is naturally more clear-cut when our starting point is contemporary atomic science. I will now attempt to specify this characteristic.

III

LET ME BEGIN BY POINTING OUT THE TENDENCY TO TIGHTEN THE PLAY OF axioms until it is reduced to the form of an alternative. This is how we can most clearly take stock of the freedom of our preliminary choice. Norman Robert Campbell, for instance, poses the initial epistemological problem in the following terms: "If the familiar laws of the electromagnetic field are true, the atom cannot consist only of electrons; and if they are not true, there is no evidence for the existence of electrons."[12] Usually, we advance only the first part of the alternative and conclude immediately that the atom, being neutral, in keeping with the laws of the electromagnetic field, must contain a positive particle. But since we are silent on the second part of the alternative, we do not recall explicitly that the *existence* of the electron was *postulated* based on the theory of the electromagnetic field. With this method of tacit simplification, we put our trust in quick realist thinking that will legitimize its conclusions in only one direction, always following the same method by which reality is legitimized through properties inherent in a substance. Campbell shows, in fact, that if we hesitate to postulate the proton that we do not isolate, we must refuse to postulate the electron that we succeed in detaching from the atom with the help of an appropriate electrical field. In other words, despite the experiments we may have carried out on the electron, we do not have the right to make of the electron something absolute. Its very existence involves a body of prior conditions. There will always be the objection that the electron is manipulated like an object in Millikan's experiment; but Millikan's experiment has no meaning beyond our conception of the electrical field. Indeed, as soon as the field is accepted, we are led to postulate the proton. On this proton, no experiment is carried out. Yet it is neither more nor less hypothetical than the electron itself. We thus readily see a correlation of hypotheses being established that go so far as to affect the *existence* of the elements that are postulated in our construction of the real. As previously pointed out, we are moving here from principles to the *thing* itself. In certain respects we can consider the electron as the object of a definition that takes on meaning thanks only to the geometrization of the electromagnetic field. In electrical science the electron is like a point that can actually receive geometric properties thanks only to postulates of belonging. We must, in a way, define an electrical affiliation of the electron to the field. And when we become more skillful and imaginative in constituting fields of varied structures, we will see the electron appear in more varied phenomena. From the outside, we will give properties to the electron, or at least new behaviors. Atoms are elements of machines that await technical refinements. They have an incalculable number of synthetic possibilities. They have yet to yield all things foreseen by principles.

In my view, it would thus be an error henceforth to consider atomistics as the analytical study of a fundamental element found at the origin of an intuition. Atomistics is, on the contrary, a wholly synthetic construction that must rely on a *body* of assumptions. That is why contemporary atomism is truly productive atomism. This atomism owes its productivity to the *compound character of the simple atom.*

This paradoxical phrase is not aimed merely at the electrical substructure discovered in the chemical atom by contemporary science. For it would surely be objected that if, from a physical point of view, the atom of a chemical substance was revealed as a complicated universe, by contrast, from the metaphysical point of view, the electron—a veritable atom—seems to have given proof of its simplicity and its oneness. At first glance, we can thus readily agree with Meyerson's observation: "The atom, as we sense perfectly well, if it is really to explain anything, must be simple,"[13] Yet, since, in a way, we deny this simplicity metaphysically, we really cannot rely on the complicated character of the chemical atom. We must, on the contrary, immediately tackle the more difficult task of showing that the electron itself—to the extent that it serves to construct atomic theory—presents an essential complexity.

First, the element must of course be assessed only within the synthesis it is charged with explaining. Specifically, it would be pointless and even unscientific to ask if the electron is simple *as such*. Rather, we might conclude that it is varied, if apprehended it in its multiple roles. But such a conclusion would be of little importance since it would be no more than a realist assertion. We must limit ourselves to assessing the electron through its functions, in phenomenal syntheses.

The complex character of a construction based on the electron is, in fact, so profound that we go so far as to accept, in our body of initial assumptions, propositions that contradict common experience. Thus, Bohr does not hesitate to assert the following at the basis of atomic physics: an electron that traces a circle around the nucleus of the atom does not emit energy, contrary to the predictions of classical electrodynamics.[14] If we reflect upon the nature and function of this proposition, we will see that it is a veritable *postulate*. It is presented in the same way as Euclid's postulate in geometry, or more accurately, in the same way as Lobachevsky's postulate. This proposition establishes, in a manner of speaking, a non-Maxwellian physics,[15] much as the negation of Euclid's postulate establishes a non-Euclidean geometry. At the same time, the complex character of the construction emerges, since we are breaking away from all the lessons about a moving electric charge garnered from common experience. If the physical consequences of the motion of the electron can be interpreted in two opposite ways, depending on whether the electron travels outside the atom or acts within it, we can no longer say that the atom contains, within itself, grounds for the physical consequences of its motion. It would then be completely pointless to posit the electron as being simple in and of itself, when this twin function is attributed to it. So, in light of this ambiguity, it does not seem very philosophical to say that the electrical nature of the atom is a *hypothesis*, nor would it be very philosophical to say that it is a *reality*. We would phrase it better, I think, by making explicit the axiomatic character of the chosen proposition. Let me be clear on this point: in electronic motion without radiation, there is obviously no question of a *real experience*; nor is there a question of a hypothesis that needs to be *verified*, since, even in a most favorable circumstance where such verification might take place, it would immediately hinder the explanation of the free electron's motion. We can therefore correctly speak only of a *postulate*. Consequently, we need not ask ourselves if that postulate corresponds to a fact, nor whether it is true. For a postulate cannot be qualified either as *real* or as *true*. It is merely the basis of a construction that alone can claim to attain a reality or a truth. But providing such an accommodation at the level of the phenomenon of first appearance still explains poorly the fate of axiomatic construction. More than any other goal, we seek, through new experiments, the coordination of thought. We must then evaluate a theory of the atom by considering a sort of pragmatism of reason, by referring to the usefulness of thought. In this respect, Bohr's postulate allowed for a powerful mathematical coordination. Henceforth, we are well founded, as a ratification of ten years of scientific history at least, to speak of Bohr's atom in the same way we speak of a Riemann surface.[16]

The geometric discontinuity of separate orbits would call for similar observations. Whereas the isolated and free electron can pass through any point in space,[17] the electron within the atom, on the other hand, is expected to follow specific trajectories while keeping away from strictly prohibited areas. The proposition that contains this prohibition can correspond neither to a positive experiment, nor even to a verifiable hypothesis. The fact that this proposition found, later on, an explanation in wave mechanics does not erase its initial epistemological character: Bohr clearly posited it as a *postulate*. Besides, the way wave mechanics assimilated Bohr's postulate quite clearly has an axiomatic air about it. Louis de Broglie's mechanics was able to demonstrate Bohr's postulate only by broadening the axiomatic base.[18] This mechanics *adds on* one more assumption; it adds a property to the electron from the outside: a wave length. It matters very little whether this *add-on* is clarified intuitively; there is no need to supplement with an image the relation of the electron to the wave length. These images practically emerge from disjointed elementary assumptions. Why should we be more demanding of the temporal character (wave frequency) than of the spatial character (form of the electron)? The theory is precisely what will constitute the link. If an image is subsequently provided, it will be solely as an illustration. Such an image would be, for example, a closed wave curve that brings to mind the arched enclosures of a vibrating string. But this image is not what made us think. It is a latecomer. Mathematical equations are primary.

It must also be noted that the point of departure for physical axiomatics can be relatively complex data. All things considered, geometry could also take a theorem as a postulate; it would be a question of choosing the first theorem carefully in order to abide by the rules of independence that any system of postulates must observe. Thus, nothing would prevent taking as a synonym of Euclid's postulate the theorem that sets at two right angles the sum of the angles of a triangle. In physics, it is not always a good idea to analyze a fact, because such an analysis can be a source of illusion, a blind obedience to poorly founded mental habits, especially when brought up in the face of a totally new experiment. Thus, it may be an illusion that urges us to analyze *geometrically* the *physical* trajectory of an electron. Our intuition of pure and simple mechanical motion eclipses our nascent and still poorly formed physical intuition of electrical phenomena. We always want the electron to be a simple carrier of a charge; we have yet to implement the a priori evaluation through synthesis that would allow us to start from a truly physical basis in order to construct a scientific physical nature. Mach already spoke, incidentally, "of physical experiments that are introduced in the same way as purely geometrical and arithmetical principles into the formal development of science."[19] In this respect he was correcting that strange opinion of Gauss, who maintained "that we can no longer contribute any essentially new principle to mechanics."[20] I believe, on the contrary, that all the sciences are renewed by a broadening of their base.

But without insisting on an attitude of mind whose faint indications are only now being perceived, let us take the problem as it was posed with Bohr's first construction. Here the rupture between the usual geometric intuition and the postulate of prescribed orbits is so sharp that it is impossible to justify this postulate a priori. It must be admitted and put to the test on the basis of the solidity of the constructions it enables. On this occasion we come upon difficulties similar to those encountered in mathematical teaching when it tries to pose, despite the usual practice, the possibility claimed by Lobachevsky of extending through a point outside a straight line two lines parallel to that straight line. Indeed, our intuition of a trajectory seems inseparable from the possibility of a continuous deformation of that trajectory. Try as we might to emphasize the fact that discontinuous forces act on the body in motion, immediate intuition always expects a continuous path to unite separate trajectories. Such a continuous path is of no interest to Bohr's method, which really puts into play only the trajectories identified a priori. Thus, this method contradicts the simplest and most fundamental intuition, that of the homogeneity of space.

Now a contradiction of such a fundamental intuition can hardly be accepted as anything but a postulate. It can hardly be brought into the argument except under cover of the freedom of axiomatic choices.

This axiomatic character of modern atomic doctrines goes so far that we prefer to bring back to the point of departure actual experiments framed by our very assumptions. Atomic physics thus goes in search of a purposely lost experiment. That is why it is the ultimate prestigious science. It causes us to think what heretofore we were limited to *seeing*. It tells us: Forget the facts that have instructed you; forget those bodies that we partition, that we dissolve, that we mingle. See this invisible world through the eyes of the mind. Contrary to a universe whose masses are stable, whose events are languid and linked, imagine a world that is multiple, discontinuous, and whose perfect mobility is without friction or kinetic wear. Just make sure, first of all, that all this is possible rationally, that is, that no inner contradiction slipped into the core of your initial assumptions. See as well that nothing superfluous was assumed, in other words, that the system of postulates is complete and quite closed. Once all these preliminary precautions have been taken, close your eyes to the real and place your trust in mental intuitions. You will build a rational world and you will produce unknown phenomena.

Will it then be said that realism is victorious in the end since we come back to reality? Will it be repeated that the chain of reasoning is no more than a scaffolding unlocking the organic character of the real? That would be to misunderstand the truly synthetic and entirely rational intuition that leads us to perceive the suitability of the initial assumptions.

It is precisely the intuition of that suitability that shapes axiomatic genius. Psychologically speaking, we did not select principles that were in fact disconnected. We merely postulated them as disconnected. In other words, modern atomic physics gives us a brilliant example of axiomatic thought. It teaches us to think about the details of atomic being as analytically independent and subsequently to show their dependence on synthesis. In particular, in no way is there a question of referring to an original simplicity. We are not, in fact, using the idea of an axiom as a synonym of a clear notion, and more than a century has elapsed between contemporary terms and those of Baudrimont, who wrote in 1833: "Geometry is true for everyone; the same can be said for atomic theory; only in order to study these two sciences, we have to rely on bases that are so simple that they slip away from demonstrations and can only be considered axioms."[21] Despite the name, we are dealing with postulates, and not axioms, in modern axiomatics. It is not because the idea of the atom is clear and simple that it is productive. It does not refer to what resists analysis but rather to what produces synthesis, so that axiomatic atomism takes its meaning only from the construction it fosters. This science lets us perceive the true nature of rationalist thought's endeavor. As Vladimir Jankélévitch so aptly states: "The interpretive effort . . . requires that the mind, when faced with problems, place itself forthwith in a spiritual attitude and *discover* real meaning by assuming it, so that theorizing will always consist, if need be, in assuming that the problem is resolved."[22] Jankélévitch also speaks "of a sort of initial venture: one has to begin, one has to take risks." Contemporary atomism is perhaps the best example of that scientific risk through which new intuitions reform thought and experiment.

CONCLUSION

IF WE CONSIDER THE MODERN VIEW OF SCIENTIFIC RESEARCH ON ATOMIC phenomena, we easily recognize the illusory character of our first intuitions. These intuitions respond too soon and too completely to the questions that are posed; they do not foster complex and productive syntheses; they do not suggest experiments. It even seems that common knowledge can be adequately characterized by its lack of a priori synthetic evaluations, by its lack of clearly stated postulates. Synthesis here is never more than a reply to analysis; it restores what analysis had disorganized. In our immediate analyses, we even forget to specify the necessarily distinctive point of view of our methods of partition. Yet it would be great progress if we could always reveal the conditions of an analysis, its limits and the points of view that determine it. We would then see that analysis is always done from a particular perspective and that the most common error is to think that substance is governed by quality. That is the original sin of realism. No doctrine has suffered from it more than atomism. Indeed, one of the most productive characteristics of positivism was to have used a qualifier to translate what belongs to *method* while leaving aside any reference to a quality that belongs to *being*. Thus, the chemical atom, for this philosophy, is nothing more than "atomic" phenomena studied by the chemical method. Whatever we may think of positivism, such a methodological perspective remains fundamental.

As one of its first obligations, the scientific method requires that we not go beyond what is defined, either in the sciences of nature or the sciences of the mind. In this regard, following Lodge's expression, the experimenter must therefore practice a veritable policy of exclusion.[1] Thus, we see how dangerous the essential assumption of realism is when it attributes to the scientific object more properties than we actually know about it. It may well be with regard to atomic science that the realist assumption is the most defective. Indeed, what we must be convinced of most of all is that *the atom is not our object*; it is not *an* object offered to our research; it is not a given; it is not a fragment of the given; it is not an aspect of the given. No intuition would thus be able to sum it up. Atomism would be better summarized by taking the atom as a center of convergence for technical methods, at the end of various processes of objectification. If, somehow, the scientific atom were suddenly to manifest itself through empirical traits, in complete disregard for technical precautions, it would indicate an instrumental failure or a methodological error. The constancy of atomic phenomena is the sign of a dependable method. In chemistry, a pure body is a conquest of a sure mind. The purity of the product is proof of the reliability of the technique.

When we have thus excluded everything that might distort experiments, we can more easily implement new techniques, consistent with mathematical ideas. The mathematical resolution of a theory thus fosters experimental precision. It is a point that Lémeray astutely highlights. "There exists," he writes, "a narrow correlation between the fact that a mathematical problem can be completely resolved and the fact that physical phenomena, constituting its concrete application, display an especially clear experimental character."[2] The scientific spirit that inspires laboratory work implements a sort of fusion of negative precautions, positive techniques, and mathematical inductions. Thus, the atom of philosophers, long a symbol of the conciliation of contradictory characteristics, makes way for the atom of physicists, the study of which brings together the most varied philosophical attitudes. This eclecticism is such that we can say that modern atomic science is clarified within all philosophical perspectives and that contemporary atomism is the most prodigious of metaphysics. Never has a swarm of ideas been so alive around things, never has our grasp of the real been prepared from so far away and by such varied means as in our conquest of the infinitely small. We would therefore be right not to neglect any of the philosophical paths that I have endeavored to retrace in the course of this work. We should even find a way to establish correspondences among the various philosophies in order to succeed in truly submitting the atom to *thought*.

With this perspective in mind, if I were to put in order (rather than condense) the philosophical attitudes that I have attempted to characterize, here is how I would conceive the philosophical teaching of atomism. First of all, I would advise adopting a *critical* approach to the issue. Indeed, it is necessary above all to recognize fully the slope along which the mind naturally and imperceptibly moves toward atomism. In its faded and artificial form, the theory developed by Hannequin is very instructive in this regard; it shows us how the mind applies discontinuity onto a continuity that is most resistant to this line of thought. This theory, however, is not rich enough in facts: never could the idea of an indivisible unit have emanated from simple geometric measurement, if experiments had always neglected everything that effectively divides up a material quantity. For a precise psychology of fragmentation, we should, after Hannequin's thesis, examine the penetrating studies of Édouard Le Roy. In following them, we would see the *simplifications* by which "separate centers, otherwise surrounded by a hazy atmosphere," are defined.[3] We would finally end up with a "schematic and formal knowledge entirely determinable by atoms apprehended by the mind."[4]

We would then recognize that these schematic atoms need to be ballasted. It is the function of simplified objects to enter complex assessments as often as possible and to take on ever greater weight. The first enrichment might, in fact, be systematically and clearly posed a priori. All that would be required would be to carry the intuition of the physical unit, a compact and resistant unit, to the very framework of a priori forms, by placing energy itself among the ranks of a priori forms. That is a modification that goes along with the way Schopenhauer rectifies Kantianism, by bringing the causality of a priori forms closer to sensibility. In this view the atom seems more a unit of cause than a unit of substance.

But, whether we like it or not, with each enrichment, we are inclined imperceptibly toward realism. From a pedagogical perspective, after a *critical* preparation, realism has the advantage, at least, of being self-aware. So it is now possible to welcome and manage the lessons of immediate experience. Such a fusion of the principles of critical atomism and the teachings of realist atomism is metaphysically impure, yet it is through this swift and frank eclecticism that we can, I think, summarize all forms of philosophical atomism. Of course we will find, from one school to the other, varying dosages that mix a little a priori with a lot of a posteriori, but never will we find the pure form. The result for the metaphysician will be an indefinable impression in the face of atomism, for we do not know if atomism discerns or proves. Looking at it more closely, we often notice, through a scandalous double inversion, that *realism seeks to prove and criticism seeks to discern.* If I insist on this mixed aspect, well suited, in fact, to give the illusion of concrete fullness, it is because such an aspect is very characteristic of philosophical atomism. That may be why most great metaphysical doctrines turned away from atomism, and why atomism thus became the symbol of opposition to the metaphysical mind.

To summarize, an initial fusion is necessary between idealist and realist theses in order to understand well the full range of philosophical atomism. If we could succeed in bringing about this fusion, we would be better prepared than we might think to follow the development of modern scientific atomism.

Let us now see what philosophical lessons we can draw from modern scientific activity in its dual experimental and theoretical forms. Specifically, in this new examination it is still necessary, I think, to try to bring together two different states of mind, with the essentially empirical position of positivism naturally preceding the constructive boldness of contemporary atomism. Besides, that is not merely a historical fact; it is a permanent pedagogical necessity. There is nothing more instructive in this regard than the entirely philosophical reaction that determined Heisenberg's investigations.[5] This reaction is a reversion to positivism that is curiously mixed with a reflection on the necessary and in some ways a priori conditions of our active observation of phenomena. Indeed, from the very start, inspired by positivism, any talk about the atom in terms that cannot be defined experimentally is rejected. Next, the atomic phenomenon is approached in its narrow correlation with the experiment that investigates it. Thus, the phenomenon is not posited as contemporaneous to the real itself, a real that is indifferent to our knowledge. On the contrary, it is a reality that has become responsive to detection and is necessarily engaged in the theoretical organization that attempts to apprehend it. Here again, two philosophies of the experiment must be associated; the theoretical construction of experiments on the atom can no longer be content with the positivist attitude, even though it actually starts from this attitude. We do not get all of theory from the experiment. Therefore we no longer worry about increasing the range of assumptions, but these assumptions are henceforth of a mathematical order. We thus hope to go beyond the mere interpretation of compensated phenomena or at least to understand the compensation itself.[6] And so is born a kind of theoretical atomism.

Furthermore, since the compensation of phenomena takes place according to the principles of probability, we find ourselves entitled to develop axiomatic positions. We will not hesitate to multiply statistical principles. Never has scientific imagination been richer, more mobile, and more subtle than in the contemporary research on atomic principles.

So we move easily from the negations characteristic of the positivist position to a priori experimental affirmations. Caution and hypotheses are intermingled. We are constantly engaged in sorting out the mathematical phenomena from both compensated and garbled physical phenomena. An active suspicion helps to amend this method of selection, for there is always the risk of forgetting variables, of deleting or adding symmetries. Far from being guided by substantialist analogies, we distrust them. Far from considering substance to be a whole, we try to break apart the solidarity of attributes. For example, as Chwolson writes: "one certainly has to notice that matter can be isotropic for one property, anisotropic for another.... Crystals of a regular system are isotropic with respect to optic properties, anisotropic with regard to elastic properties."[7] Hence the need to multiply points of view, to go to the infinitely small through a plurality of paths, by surrounding it with an intricate network of theorems.

By contrast, classical positivism would lead us to overvalue certain facts excessively, to consider the real through only one of its attributes. Positivism learns, in effect, through contact with the immediate phenomenon; it is inclined to take the immediate phenomenon for the important phenomenon, for the only phenomenon able to authorize theory. But modern science has reconciled us with the causality of the infinitely small, with the geometry of detail. It has often been said that in chemistry discoveries were made by studying the residues discarded by coarse experiments. We could say that contemporary atomism is likewise found in the residues discarded by immediate positivism. Thus does the science of the atom complete chemistry through geometry. Intuitions of the senses must give way to rational intuitions. And finally, if philosophical thought were one day to fill the gap that separates naïve atomism from contemporary scientific atomism, the same question would still require an answer: How can intuitions of the senses become, little by little, rational intuitions; how can facts help to discover laws; and especially, how can laws become so strongly organized as to suggest rules?

NOTES

NOTES TO INTRODUCTION

1. [A categorem is a term capable of standing alone as the subject or predicate of a logical proposition. In this instance Bachelard points to an expansion of the idea of the atom over time, with the result that it functions like a categorematic expression broad enough to encompass contradictory notions.]

2. L[éon] Brunschvicg, *L'éxpérience humaine de la causalité physique* (Paris: Alcan, 1922), 381. [Léon Brunschvicg (1869–1944), French idealist philosopher and co-founder of the *Revue de métaphysique et de morale* directed one of Bachelard's two doctoral dissertations.]

3. [Émile Bréhier (1876–1952) was a French historian of philosophy. His two-part, seven-volume *Histoire de la philosophie* (Paris: Alcan, 1928; Paris: Presses Universitaires de France, 2012) covers the history of philosophy from antiquity and the middle ages to modern times. Hereafter cited as *Histoire de la philosophie*.]

4. It is perhaps inconceivable that [freedom] be constructed, deduced, or even proved except by experiencing it. It contradicts all attempts at coordination. "Nowhere has the idol of elucidation brought about more insoluble aporias than with questions pertaining to freedom." Vladimir Jankélévitch, *Revue de métaphysique et de morale* (December 1928): 457.

5. [From the Latin for incline, slope, or turning aside, *clinamen* is a term Lucretius applied to the unpredictable swerve of atoms, as a means of justifying the atomistic doctrine of Epicurus.]

6. K[urd] Lasswitz, *Atomistik und Kriticismus* (Braunschweig: [Vieweg und Sohn,] 1878), 49. (French Editor's note). [Hereafter cited as *Atomistik und Kriticismus*.]

7. [Charles Renouvier, cited in] L[éopold] Mabilleau, *Histoire de la philosophie atomistique* (Paris: Alcan, 1895), 52[–53. Mabilleau hereafter cited as *Histoire de la philosophie atomistique*].

8. L[ouis] Weber has shown the idealist character of modern atomism. See *Vers le positivisme absolu par l'idéalisme* (Paris: Alcan, 1903), 24ff.

NOTES TO CHAPTER I

1. See Descartes, *Oeuvres complètes*, ed. Adam-Tannery, vol. 12 [(Paris: Léopold Cerf, 1903)] *Vie de Descartes*, 17, note. [Hereafter cited as *Oeuvres complètes*.]

2. [Lasswitz, *Atomistik und Kriticismus*.]

3. [Ernst Mach, 1838–1916, Austrian physicist and philosopher after whom is named the number that identifies the ratio of a given speed to sound.]

4. See [Gaston Bachelard,] *Le pluralisme cohérent de la chimie moderne* (Paris: Vrin, 1932). [Hereafter cited as *Le pluralisme cohérent*. For further discussion, see also Roch C. Smith, *Gaston Bachelard: Philosopher of Science and Imagination* (Albany: State University of New York Press, 2016), 19–21. Hereafter cited as *Philosopher of Science and Imagination*.]

5. [Bachelard frequently challenges the ideas of French philosopher Henri Bergson (1859–1941). See Jean-François Perraudin, "Bachelard's 'Non-Bergsonism,'" in *Adventures in Phenomenology: Gaston Bachelard*, ed. Eileen Rizo-Patron, Edward S. Casey, and Jason Wirth (Albany: State University of New York Press, 2017), 29–47.]

6. [A mesomorphic state ("état mésomorphe") is defined as a state of matter whose symmetry is intermediate between a solid and a liquid and that is found especially in certain elongated organic molecules (liquid crystal) "état de la matière dont la symétrie est intermédiaire entre celle d'un solide et celle d'un liquide et qui se rencontre surtout avec certaines molécules organiques allongées (cristal liquide)." http://www.larousse. fr/dictionnaires/français.]

7. A[uguste] Lumière, *Théorie colloidale de la biologie et de la pathologie* (Paris: Étienne Chiron, 1922), 69.

8. J[ean]-A[ndré] Deluc, *Lettres physiques et morales sur l'histoire de la terre et de l'homme*, vol. 2 (Paris: V. Duchesne, 1780), 29.

9. Ibid., 30.

10. [Dr. Edmond Locard, Bachelard's contemporary, established the basic principle of forensic science that "every contact leaves a trace," known as Locard's exchange principle. https://sites.google.com/site/apchemprojectforensicchemistry/experts-in-the-field.]

11. H[élène] Metzger, *Les doctrines chimiques en France du début du XVIIe à la fin du XVIIIe siècle* (Paris: P[resses] U[niversitaires de] F[rance], 1923), 61. [The "dabbler in pharmacy" mentioned by Metzger is E. R. Arnaud, a Doctor of Medicine who published an *Introduction à la chymie ou à la vraye physique* in 1656, where he sought to improve medicine by "having the physician enter the pharmacist's laboratory to discover … that the sciences, far from combatting each other, must strive for mutual enlightenment" ("faire pénétrer le médecin dans le laboratoire du pharmacien, lui

montrer... que les sciences, loin de se combattre, doivent essayer de s'éclairer mutuel-lement") (Metzger, 59–60). Hereafter cited as *Les doctrines chimiques en France.*]

12. Ibid., 372.

13. [The reference here is to the famed work of the French Enlightenment, the eighteenth century *Encyclopédie*, edited by Denis Diderot and Jean le Rond d'Alembert.]

14. L[éon] Robin, *La pensée grecque et les origines de l'esprit scientifique* (Paris: Albin Michel, 1923), 82. See also 145. [Hereafter cited as *La pensée grecque.*]

15. Bréhier, *Histoire de la philosophie*, 1: 80. Bréhier refers to *Lucrèce*, I, 370.

16. [With this "sort of passage to the limit" Bachelard seems to be anticipating the calculation to the limit that will be used in calculus to measure the infinitesimal.]

17. Descartes, *Oeuvres complètes*, 6: 240.

18. Aristotle, *Physique*, bk. 4, chap. 8, §9, French trans. B[arthélemy]-Saint-Hilaire, 191; [Paris: Vrin, French trans. A. Stevens, 2012 (Editor's note)]. [For English trans., see Aristotle, *Physics*, bk. 4, chap. 6 (213b, 23–27).]

19. Ibid., bk. 4, chap. 9, §18. [English trans., bk. 4, chap. 8 (216a, 20).]

20. Robin, *La pensée grecque*, 337. [The 2015 printing of *Les intuitions atomistiques*, on which this translation is based, mistakenly attributes this citation to Aristotle. The quotation, as indicated in the original 1933 edition, and as noted here, is from Robin.]

21. B[arthélemy]-Saint-Hilaire, preface to Aristotle's *Physique*, xiv.

22. This observation continues to be related in Diderot and d'Alembert's *Ency-clopédie*, s.v. "Vuide," last column.

NOTES TO CHAPTER II

1. Robin, *La pensée grecque*, 136[–137]. [The italics indicate Robin's translation of Greek terms that are not included in Bachelard's long quotation. It should be noted that the 2015 printing of *Les intuitions atomistiques*, on which this translation is based, unlike the original 1933 work, and Robin's own text, mistakenly closes the Robin quotation after the first sentence of the quoted page. I have followed the 1933 version in this instance.]

2. Mabilleau, *Histoire de la philosophie atomistique*, 39.

3. Robin, *La pensée grecque*, 137.

4. Mabilleau, *Histoire de la philosophie atomistique*, 189.

5. [See Bachelard's discussion of "positivist atomism" in chap. 4.]

6. Mabilleau, *Histoire de la philosophie atomistique*, 534.

7. A[rthur] Hannequin, *Essai critique sur l'hypothèse des atomes dans la science contemporaine* (Paris: G. Masson, 1895), 245. [Hereafter cited as *L'hypothèse des atomes.*]

8. See, for example, P[ierre] Gassendi, *Syntagma philosophicum* (1658), part 1, bk. 5, chaps. 9, 10, and 11.

9. [Nicolas Lémery (1645–1715), French chemist.]

10. P[ierre] Duhem, *Le mixte et la combinaison chimique: essai sur l'évolution d'une idée* (Paris: C. Naud, 1902), 20–21.

11. Nicolas Lémery, *Cours de chymie*, contenant la manière de faire des opérations qui sont en usage dans la médecine, par une méthode facile.... Nouvelle édition revue et corrigée par M. Baron (*Course in Chemistry*, including the means of doing operations utilized in medicine, through an easy method .... New edition, revised and corrected by M. Baron) (Paris: Laurent-Charly D'Houry, 1757), 18–19.

12. Metzger, *Les doctrines chimiques en France*, 433.

13. [Wilhelm (Guillaume) Homberg (1652–1715), Dutch scientist and physician.]

14. *Histoire et mémoires de l'académie des sciences*, 1706, 262. Quoted by H[élène] Metzger in *Les doctrines chimiques en France*, 380.

15. [Opium makes you sleep because it has soporific qualities, "explains" the doctor in Molière's *Le malade imaginaire* (*The Imaginary Invalid*).]

16. Voltaire, *Dictionnaire philosophique*, s.v. "Atomes."

17. See M[arcellin] Berthelot, *La synthèse chimique* (Paris: Germer Baillière, 1876), 33. [Hereafter cited as *La synthèse chimique*.] [An antonym of "heterogeneous," "homeomerous" refers to the principle put forth by Greek philosopher Anaxagoras (fifth century BCE) that large entities are made up of identical smaller entities. For a fuller discussion, see Patricia Curd, "Anaxagoras," in *The Stanford Encyclopedia of Philosophy* (Winter 2015), ed. Edward N. Zalta, https://plato.stanford.edu/archives/win2015/entries/anaxagoras.]

18. A[rthur] Schopenhauer, *Le monde comme volonté et représentation*, French trans. A. Burdeau, vol. 2 (Paris: [Alcan], 1844), 52.

19. [Augustin-Louis Cauchy, French mathematician, 1789–1857.]

20. Lasswitz, *Atomistik und Ktiticismus*, 49.

21. Mabilleau, *Histoire de la philosophie atomistique*, 277.

22. Bréhier, *Histoire de la philosophie*, 2: 343.

23. I[saac] Todhunter, *A History of the Theory of Elasticity*, vol. 1 ([Cambridge: Cambridge] University Press, 1886), 4.

24. [Sir] William Petty, 1674: *Letters to John Aubrey*. [For a recent edition by Thomas E. Jordan, see Lewiston, NY: Edwin Mellen Press, 2010.]

25. A[lexandre] Koyré, *La philosophie de J. Boehme* (Paris: Vrin, 1929), 131.

26. F[rançois] Pillon, ed., *Année philosophique*, 1891 [(Paris: Alcan, 1892)], 70. [Hereafter cited as *Année philosophique*.]

27. Ibid., 71. [Nicolas Malebranche (1638–1715) "is known for his occasionalism, that is, his doctrine that God is the only causal agent, and that creatures merely provide

the 'occasion' for divine action." Tad Schmaltz, "Nicolas Malebranche," in *The Stanford Encyclopedia of Philosophy* (Winter 2016), ed. Edward N. Zalta, https://plato.stanford. edu/archives/win2016/entries/malebranche/.]

28. See K[urd] Lasswitz, *Geschichte der Atomistik*, 2nd ed., vol. 2 (Leipzig: Voss, 1926), 41.

29. J[oseph] Prost, *Essai sur l'atomisme et l'occasionnalsime dans la philosophie cartésienne* (Paris: H. Paulin, 1907), 56.

30. Pillon, *Année philosophique*, 98.

31. Ibid., 100.

32. H[élène] Metzger, *Newton, Stahl, Boerhaave et la doctrine chimique* (Paris: Alcan, 1930), 37.

33. Ibid., 77–78.

34. Pillon, *Année philosophique*, 99.

35. [Roger Joseph Boscovich (1711–1787), Croatian physicist and mathematician.]

36. Nikola M. Poppovich, *Die Lehre vom diskreten Raum* (Wien: Braunmüller, 1922). [Augustin-Louis Cauchy, see n. 19; Johan Friedrich Herbart (1776–1841), German philosopher, mathematician, and psychologist; Charles Bernard Renouvier (1815–1903), French philosopher; François Evellin (1835–1910), French philosopher.]

37. [Branislav Petronievics (1875–1954), Serbian philosopher.]

NOTES TO CHAPTER III

1. P[aul] Kirchberger, *La théorie atomique: son histoire et son développement*, French trans. M. Thiers (Paris: Payot, 1930), 18. [Hereafter cited as *La théorie atomique*.]

2. See Lasswitz, *Atomistik und Kriticismus*, 32. [The references are to Hermann von Helmholtz (1821–1894) and Sir William Thomson, Lord Kelvin (1842–1902).]

3. Marcelin Berthelot, *La synthèse chimique*, 7.

4. Ibid.

5. Kirchberger, *La théorie atomique*, 10. [Hans Vaihinger (1852–1933), German philosopher; see also chap. 4, n. 4; Otto Lehmann (1855–1922), German physicist; Andreas von Antropoff (1878–1956), German chemist.]

6. [Marcelin Berthelot, *La synthèse chimique*, 34.]

7. J[ustus] von Liebig, *Nouvelles lettres sur la chimie*, ed. C[harles] Gerhardt, vol. 2 (Paris: V. Masson, 1852), 292. [Inflammable air we now call hydrogen and dephlogisticated air is now known as oxygen. See https://www..britannica.com/biography/Henry Cavendish.]

8. [Bachelard is referring to *Le pluralisme cohérent de la chimie moderne*, published the previous year in 1932. See 46–56. Claude Louis Berthollet (1748–1822) and Joseph

Louis Proust (1754–1826) were both French chemists who engaged in a long-running disagreement over the way chemical bodies combine. For a detailed discussion, see, for example: Kiyohisa Fuji, "The Berthollet-Proust Controversy and Dalton's Chemical Atomic Theory, 1800–1820," *The British Journal for the History of Science* 19, no. 2 (July 1986): 177–200.]

9. Quoted by R[aoul] Jagnaux, *Histoire de la chimie*, vol. 2 (Paris: Baudry et Cie, 1891), 227.

10. Sterry Hunt, *Un système chimique nouveau* [French trans. Walter Spring] (Paris-Liège: G. Carré-M. Nierstrasz, 1889), 16.

11. [Ibid., xvi. Walter Spring (1848–1911), French chemist.]

12. Ibid., 54.

13. [Allotropy is defined as "the existence of a chemical element in two or more forms. . . . . Elements exhibiting allotropy include tin, carbon, sulfur, phosphorous, and oxygen." https://www.britannica.com/science/allotropy.]

14. Daniel Berthelot, *De l'allotropie des corps simples* (Paris: G. Steinheil, 1894), 3. [Hereafter cited as *Allotropie*.]

15. [It should be evident that Bachelard refers here to "phenomenology" as the empirical manifestation of phenomena and not in the philosophical sense that will mark his later works such as *The Poetics of Space* (French publication, 1957), *The Poetics of Reverie* (French publication, 1960), and the posthumous *Fragments of a Poetics of Fire* (French publication, 1988).]

16. Daniel Berthelot, *Allotropie*, 3. [Berthelot refers to Jöns Jacob Berzelius (1779–1848), a Swedish chemist who established the law of constant proportions. He is considered one the founders of modern chemistry. www.chemheritage.org.]

17. [Marcelin Berthelot (1827–1907) and his son Daniel Berthelot (1865–1927) were French organic and physical chemists.]

18. Quoted by M[aurice] Meslans, *États allotropiques des corps simples* (Paris: [Georges Carré,] 1894), 35.

19. [Bachelard gives no reference for this quote, however Marcelin Berthelot discusses the transformation brought to formene (marsh gas) and benzine by extreme heat in *La synthèse chimique* (Paris: Germer Baillière, 1876).]

20. M[arcel] Boll, *Cours de chimie [à l'usage des candidats aux grandes écoles* (Paris: Dunod)], (1920), 51.

21. [René Antoine Ferchault de Réaumur], *Mémoires de l'académie* (Paris: [Imprimerie royale,] 1711), [6 and 81]. [Réaumur (1683–1757) was a French scientist, principally an entymologist.]

22. G[eorges] Urbain, *Les disciplines d'une science* (Paris: G. Doin, 1921), 316.

23. Ibid., 321.

24. From an experimental point of view one can consult the works of Paul Pascal on additive properties.

NOTES TO CHAPTER IV

1. Kirchberger, *La théorie atomique*, 49. [Auguste Comte (1798–1857) is widely recognized as the founder of positivism in France. For more on Jöns Jakob Berzelius, *see* chap. 3, n. 16.]

2. See the origins of this tendency in the renowned treatises of L[eopold] Gmelin, 1788–1853. [German chemist, author of the multivolume *Handbook of Chemistry*.]

3. M[aurice] Delacre, *Essai de philosophie chimique* (Paris: Payot, 1923), 35. [Hereafter cited as *Philosophie chimique*.]

4. H[ans] Vaihinger, *Die Philosophie des Als Ob* (Leipzig: Meiner, 1922), 103. [Hereafter cited as *Die Philosophie des Als Ob*.]

5. [John Dalton (1766–1844), English chemist.]

6. [Jeremias Richter (1762–1807) established stoichiometry as "the science of measuring the quantitative proportions or mass ratios in which chemical elements strand to one another." http://www.encyclopedia.com/science/dictionaries-thesauruses-pictures-and-press-releases/richter-jeremias-benjamin.]

7. F[riedrich]-A[lbert] Lange, *Histoire du materialisme*, French trans. B. Pommerol, vol. 2 (Paris: A. Costes, 1921), 195. [Hereafter cited as *Histoire du materialisme*.]

8. Delacre, *Philosophie chimique*, 43.

9. G[eorges] Urbain, *Les notions fondamentales d'élément chimique et d'atome* (Paris: Gauthier-Villars, 1925), 19–21.

10. [Ibid., 21–22.]

11. P[aul] Schutzenberger, *Traité de chimie générale*, 2nd ed. (Paris: Hachette, 1884), introduction, v.

12. ["Phenomenology" is used here and elsewhere in this chapter as the empirical display of phenomena. See also chap. 3, n. 15.]

13. [See chap. 3, n. 8.]

14. [See chap. 3, n. 16.]

15. [Bachelard will return to the role of large numbers when discussing mathematical "compensation" in his conclusion. See p. 100 and p. 114 n6.]

16. [The Dulong-Petit law, discussed here by Bachelard, was presented to the French Academy of Sciences on April 12, 1819, by French physicists Pierre-Louis Dulong (1758–1839) and Alexis Thérèse Petit (1791–1820). Their findings were then published as "Sur quelques points importants de la théorie de la chaleur" (On several important

points regarding heat theory) in *Annales de chimie et de physique*, vol. 10 (Paris: Gro-
chard, 1819), 395–413.]

  17. [Ibid., 395–396.]

  18. [François-Marie Raoult (1830–1901), French chemist. Raoult's findings on
lowering freezing points of solutions were published between 1878 and 1886. See the
research report: "Notes sur les publications de F-M Raoult. Cryoscopie," in *Les classiques
de la science*, vol. 4 (Grenoble: [University Press], 1886), 4–7.]

  19. [Amadeo Avogadro (1776–1856), Italian chemist; Joseph Louis Gay-Lussac
(1778–1850), French chemist. For a discussion of Avogadro's law in relation to the work
of Gay-Lussac, see https://www.chemheritage.org/historical-profile/amedeo-avogadro.
See also https://www.chemheritage.org/historical-profile/joseph-louis-gay-lussac.]

  20. [Avogadro's findings, quoted here by Bachelard without bibliographical reference,
are cited in Henry Le Chatelier, ed., *Molecules, atomes, et notations chimiques* (Paris:
Armand Colin, 1913), 17.]

  21. [Jean-Baptiste-André Dumas, "Sur quelques points de la théorie atomistique,"
ibid., 37. Dumas (1800–1884) was a French chemist.]

  22. [Jean-Baptiste Perrin (1870–1942). "French physicist who, in his studies of the
Brownian motion of minute particles suspended in liquids, verified Albert Einstein's
explanation of this phenomenon and thereby confirmed the atomic nature of matter."
https://www.britannica.com/biography/Jean-Perrin.]

  23. [Nowadays a more commonly given equivalent for Avogadro's number is 6×10^{23}
or 6.022×10^{23}.]

NOTES TO CHAPTER V

  1. [Bachelard's italics. Elsewhere in this chapter and beyond, to avoid confusion with
common usage, I have italicized words like *criticism* and *critical* in order to underline
the Kantian connotations. Bachelard's occasional italics for these words are identified
as such. Expressions such as "critical theory," "critical doctrine," "critical philosophy,"
"critical thought," and "atomistic criticism," whose Kantian allusions are evident, are
left unitalicized.]

  2. Hannequin, *L'hypothèse des atomes*, 3.

  3. [Bachelard's italics, here and in the next sentence.]

  4. Hannequin, *L'hypothèse des atomes*, 26.

  5. Ibid., 12.

  6. Ibid., 26.

  7. Ibid., 69.

  8. Ibid., 11.

9. Ibid., 34.

10. Ibid., 36.

11. [Bonaventura Cavalieri (1598–1647), Italian mathematician. His work on "indivisibles" was a precursor to integral calculus. See www.britannica.com/biography/Bonaventura-Cavalieri.]

12. Hannequin, *L'hypothèse des atomes*, [36–]37.

13. [Louis Couturat], *Revue de métaphysique et de morale* (1896–1897). [Bachelard does not provide page references, but Couturat's series of three articles on Hannequin can be found in vol. 4 (1896): 778–797, and in vol. 5 (1897): 87–113 and 220–247.]

14. Hannequin, *L'hypothèse des atomes*, 70[–71]. See also Lange, *Histoire du matérialisme*, 2: 191.

15. [Hannequin, *L'hypothèse des atomes*, 73. In several instances, especially when quoting Hannequin, Bachelard relies on the context for references to cited material. In the interest of clarity, here and elsewhere, such references are provided in brackets by the translator.]

16. Ibid.

17. [Ibid.]

18. [A Kantian term used by Hannequin, phoronomy refers to motion as a trajectory, without reference to the laws of force or mass. See the subsequent discussion by Bachelard.]

19. Hannequin, *L'hypothèse des atomes*, 81.

20. Ibid., 82.

21. Ibid.

22. [Ibid.; Bachelard's italics.]

23. [Ibid., 83.]

24. Ibid., 90.

25. [Ibid.]

26. [Ibid., 91.]

27. Ibid., 91.

28. See [Gaston Bachelard], *La valeur inductive de la relativité*, chap. 3 (Paris: Vrin, 2014 [1929]).

29. Hannequin, *L'hypothèse des atomes*, 92.

30. [Lasswitz, *Atomistik und Kriticismus*, v.]

31. Ibid., 6.

32. Lasswitz refers back to Kant, *Prolegomena zu einer künftigen Metaphysik* (Riga: [J. F. Hartknoch,] 1783), S. 111; French trans. L. Guillermit (Paris: Vrin, 1997). [See also *Prolegomena for Any Future Metaphysics*, trans. L. W. Beck (Indianapolis: Bobbs-Merrill, 1950).]

33. [The italicized comments in parentheses are Bachelard's. My translation of Bachelard's "science de l'expérience" and "science de la connaissance" as "empirical

science" and "science of empirical knowledge" is in keeping with the term "Erfahrung-swissenschaft" used by Lasswitz and cited here by Bachelard.]

34. Lasswitz, *Atomistik und Kriticismus,* 43[–44].

35. Ibid., 52.

36. Ibid., 57.

37. Ibid., 62.

38. [Bachelard's italics.]

NOTES TO CHAPTER VI

1. ["To know is to describe in order to retrieve," writes Bachelard at the beginning of his very first book, *Essai sur la connaissance approchée* (An Essay on Knowledge by Approximation) (Paris: Vrin 1973 [1928]), 9. Description and retrieval anchor much of Bachelard's subsequent epistemological discussion. Here Bachelard insists on the modified and restricted role definition plays within the synthetic (as opposed to analytic) description of axiomatic atomism. See also p. 87–88. See also Smith, *Philosopher of Science and Imagination,* 9–15.]

2. A[ndré] Lalande, *Les théories de l'induction et de l'expérimentation* (Paris: Boivin et Cie, 1929), 146ff.

3. See Vaihinger, *Die Philosophie des Als Ob,* 150. "Das Atom ist keine naturwissenschaftliche *Entdeckung,* sondern eine Erfindung" [The atom is not a natural scientific *discovery* but an invention].

4. Lasswitz, *Atomistik und Kriticismus,* 12.

5. Ibid., 33.

6. Ibid., 41.

7. [Bachelard was writing in 1933 of course, but this continues to be true today, where elements still do not carry the names of those who first discovered them. While certain chemical elements have been named after individuals, such as Rutherfordium (after New Zealand physicist Ernest Rutherford), Seaborgium (after American nuclear chemist Glenn T. Seaborg), Bohrium (after Danish physicist Niels Bohr), and Roentgenium (after German physicist Wilhelm Conrad Röntgen), all were discovered through the work of more recent scientists and were named in honor of these predecessors. My thanks to Professor Milton S. Feather for leading me to this information. Pieter Zeeman (1865–1943] was a Dutch physicist; Johannes Stark (1874–1957) was a German physicist; Arthur Holly Compton (1892–1962) was an American physicist; Chandrasekhara Venkata Raman (1888–1970) was an Indian physicist.]

8. [Claude Bernard (1813–1878), French physiologist who is widely acknowledged for his contributions to the scientific method.]

9. C[laude] Bernard, *Introduction à la médecine expérimentale* (Paris: Baillière, 1865), 48–49.

10. [Robert Millikan (1868–1953), American physicist; Otto Stern (1888–1969), German American physicist; Walter Gerlach (1889–1979), German physicist. Millikan developed a device to directly measure, through an "oil-drop experiment," the electric charge of an electron. Stern and Gerlach devised an instrument that demonstrated the quantum properties of atoms. For further reading, see https://www. britannica. com/science/Millikan-oil-drop-experiement and https://britannica.com/science/ quantum-mechanics-physics.]

11. Vaihinger, *Die Philosophie des Als Ob*. See especially the chapters on *Das Atom als Fiktion* [The Atom as Fiction] and *Die Atomistik als Fiktion* [Atomism as Fiction].

12. [Bachelard quotes the French translation:] N[orman] R[obert] Campbell, *La structure de l'atome*, French trans. A. Corvisy (Paris: Barnéoud, 1925), 1. [In the English original Campbell writes: "the atom could not consist of electrons only, if the familiar laws of the electromagnetic field were true; and if they were not true, there is no evidence for the existence of electrons." *Modern Electrical Theory, Supplementary Chapters*, chap.17, "The Structure of the Atom" (Cambridge: Cambridge University Press, 1923), 1.]

13. É[mile] Meyerson, *Identité et réalité* (Paris: Alcan, 1912), 67.

14. [Danish physicist Niels Bohr (1885–1962) proposed a model of the atom in 1913 in which electrons were limited to prescribed orbits. Electrons emitted radiation, in the form of quantum light, only when they jumped from one orbit to another. For further discussion, see https://www. britannica.com/science/Bohr-atomic-model.]

15. [Bachelard's use of the phrase "non-Maxwellian physics" in this context refers to electromagnetic theory and the idea that the Bohr electron does not radiate electromagnetic radiation as Maxwell's Laws specify for a moving charge. It should not be mistaken for a reference to the Maxwell-Boltzmann distribution associated with thermodynamic physics. (My thanks to Professor Stephen Danford for this observation.) Nikolai Lobachevsky (1792–1856) was a Russian mathematician and a founder of non-Euclidian geometry.]

16. [Bernard Riemann (1826–1866), German mathematician. A Riemann surface has been described as "a one-dimensional complex manifold [that] can be thought of as [a] 'deformed version' of the complex plane" (Yahlee.info/what-is-rieman-surface. html). For an illustration, see http://3d-xplormath.org.]

17. It might be pointed out, as well, that this assumption also relates to the definition of freedom of motion. We cannot, through an experiment, make this proposition specific enough to be declared a fact.

18. [Louis de Broglie (1892–1987) was a French physicist who predicted the wave nature of electrons. See https://www.britannica.com/biography/Louis-de-Broglie.]

19. E[rnst] Mach, *La mécanique*, French translation, E. Picard (Paris: Hermann, 1904), 254. [See also Bachelard's chapter 1, note 3.]

20. Quoted by A[bel] Rey, in *L'énergétique et le mécanisme* (Paris: Alcan, 1908), 28. [Karl Friedrich Gauss (1777–1885), German mathematician.]

21. A[lexandre] Baudrimont, *Introduction à l'étude de la chimie par la théorie atomique* (Paris: Louis Colas, 1833), 102.

22. Vladimir Jankélévitch, *Revue de métaphysique et de morale* (December 1928): 465.

NOTES TO CONCLUSION

1. [Bachelard does not provide Lodge's first name, but it is likely he is referring to Sir Oliver Lodge (1851–1940), a British physicist, some thirty years Bachelard's senior, who wrote extensively on the history of physics. See, for example, Peter Rowlands, "Sir Oliver Lodge and Relativity," *Institute of Physics, History of Physics Group Newsletter* 18 (Summer 2005): 19–29, https://www.iop.org/activity/groups/subject/hp/newsletter/archive/file_66466.pdf.]

2. E[rnest]-M[aurice] Lémeray, *Leçons élémentaires sur la gravitation* (Paris: Gauthier-Villars, 1921), 5.

3. [Édouard Le Roy] *Revue de métaphysique et de morale* (1899): 381.

4. Ibid., 539.

5. [As Daniel Parrochia points out in his preface (note 12), this sole mention of Werner Heisenberg (1901–1976) can be explained by the recency of quantum physics at the time Bachelard was writing.]

6. [In mathematics the law of large numbers, sometimes referred to as *the law of compensation*, guarantees that experimental probability will more closely align with theoretical probability as more experiments are done and data is collected. Such *compensation* or application of the law of large numbers to "garbled physical phenomena" leads, as Bachelard points out, to an axiomatic rationalism that extends our empirically derived understanding and opens the way to further research.]

7. O[rest] D[anilovich] Chwolson, *Traité de physique*, French trans. E. Davaux, vol. 1 (Paris: A. Hermann, 1908), 30.

INDEX

acceleration, 76

Adam, Charles, 13, 14

additivity, 49, 50

air, 19; as wind, 25; dephlogisticated, 45. *See also* dust

air pump, 25

alchemists, 19

algebra, 90; and analysis (mathematical), 69

allotropy, 47–48. *See also* chemistry

analysis/analytical, 1, 2, 6, 9, 16, 31, 89; absolute , 23; chemical, 58; geometrical, 14, 94; (mathematical), 69–73, 78; statistical, 25; and synthesis, 2, 3, 5, 6, 15, 44, 60, 88, 92, 96, 97. *See also* algebra; atomism; calculus.

Anaxagoras, 34

ancient(s), 24, 25; as philosophers, 57; and physics, 40; and realism, 48. *See also* atomism: of past centuries

antiquity, 3, 6, 37, 45; physical science of, 5

Antropoff, Andreas von, 44

approximation, 50, 62, 63, 64, 72

arbitrariness/arbitrary, 7, 63, 78, 87

architectonics: of the science of the atom, 87

arithmetic/arithmetical, 49, 50, 62, 63, 70, 94

Aristotle, 22, 23–24, 27, 38, 39, 40

Arnaud, E. R., 21

atom(s): and absolute properties, 81; as cause, 34–35, 74–76, 99; of chemists, 67–68; compound character of, 92; Democritean, 29; and duality, 7; electrical properties of, 55; electrical substructure of, 92; and experience. 29; framework of, 87; freedom of, 5, 35–36; of internal realism, 29, 36; as logical unit, 28; naïve, 32, 37; not self-sufficient, 3; nucleus of, 92; and number, 68; of philosophers, 98; of physicists, 67–68, 98; physics of, 22; punctiform, 83; as a role, 72; size of, 80–81; submitted to thought, 98; as word, 7, 56, 86

atomic/atomistic hypothesis, 1, 55, 56, 62, 63, 90, 91

atomism: and analysis (mathematical), 69, 70–71, 72, 73; atomistic alternative, 67; contemporary, 90, 92, 96, 98, 100, 101; dogmatism of, 79; and duality, 7, 8; of extension, 74; externally realistic, 29; geometrical, 77–78; and history, 7, 79; as hypothesis, 68; learned, 29; after Lucretius, 6; materialism and, 14, 31; mechanical, 77, 78;